Radiative Transfer
in Moving Media

Springer
*Singapore
Berlin
Heidelberg
New York
Barcelona
Budapest
Hong Kong
London
Milan
Paris
Santa Clara
Tokyo*

Radiative Transfer in Moving Media

Basic Mathematical Methods
for Radiative Transfer
in Spherically Symmetrical Moving Media

K K Sen and S J Wilson

K.K Sen
Formerly Professor of Mathematics
University of Singapore

S.J. Wilson
Professor of Mathematics
Depatment of Mathematics
National University of Singapore
10 Kent Ridge Crescent
Singapore 119260

Library of Congress Cataloging-in-Publication Data

Sen, K. K., 1922-
   Radiative transfer in moving media : basic mathematical methods
for radiative transfer in spherically symmetrical moving media /
K. K. Sen, S. J. Wilson.
       p.   cm.
   Includes bibliographical references and index.
   ISBN 9813083123
       1. Stars--Atmospheres--Mathematics.   2. Radiative transfer-
-Mathematics.   3. Astrophysics.   I.Wilson, S. J.   II. Title.
   QB809.S46   1997
   523.8--dc21   97-23978
                                                                  CIP

ISBN 981-3083-12-3

This work is subject to copyright. All rights are reserved, whether the whole or part
of the material is concerned, specifically the rights of translation, reprinting, re-use
of illustrations, recitation, broadcasting, reproduction on microfilms or in any
other way, and storage in databanks or in any system now known or to be invented.
Permission for use must always be obtained from the publisher in writing.

© Springer-Verlag Singapore Pte. Ltd. 1998
Printed in Singapore

The publisher makes no representation, express or implied, with regard to the
accuracy of the information contained in this book and cannot accept any legal
responsibility or liability for any errors or omissions that may be made.

Typesetting: Camera-ready by authors
SPIN 10636390   5 4 3 2 1 0

Preface

In the first quarter of the twentieth century, spectral observations of stars revealed that some of them had Doppler shifts far in excess of that expected from thermal motion alone. They exhibited unusual broadening of spectral lines — blue shift of absorption and red shift of emission lines. Included in this category of stars are super-giants, WR-, Of-, Be-, P- cygni and protostars. All of these are surrounded by extensive, spherically symmetric moving atmospheres. To study the problems of radiative transfer and line formation in them, the transfer equations are usually expressed in spherical coordinates. The motions involved in such stellar atmospheres are expansion, rotation or a combination of both, often attended with outflow of matter from their surfaces.

In the early years, Struve (1929) attributed these abnormal spectral structures to the existence of non-thermal turbulence in the atmospheres. Beals (1929–1950) assumed that these were due to the expansion of their atmospheres with sonic and supersonic velocities, at times attended with accretion of matter from the surface. His main interest was to study the spectral structures of WR-stars. The atmosphere around them was assumed to be plane parallel. The excessive broadening of the spectral lines was explained by classical Doppler shifts (including aberration and advection) caused by the presence of velocity field in the medium. Chandrasekhar (1934), Gerasimovič (1934) and Wilson (1934) worked out the theory of radiative transfer in such stars with moving envelopes. They considered the envelope to be scattering coherently, optically thin and plane parallel. The velocity law was assumed to be linear. The abnormal spectral features were taken to be due to the Doppler shift. The influences of aberration and advection were ignored in their analysis.

The advancement of observational techniques over the years led to the discovery of a large number of stars belonging to these classes, exhibiting the complex spectral structures. There was an urgent need to search for new ideas and techniques to understand and tackle these problems. This led to the introduction of the concept of "photon escape probability" and the wisdom of expressing the transfer equations in "comoving frames" (CMF). The new concept of escape probability was proposed by Sobolev (1960). He asserted that the presence of large scale velocity gradients in the stellar atmospheres offers an escape mechanism for the photons to cross the outer surface of the medium without actually interacting with the medium. He did not go through the conventional process of solving the transfer equation. The technique of using comoving frame transfer equation was first proposed by McCrea and Mitra (1936). It was solved by Chandrasekhar (1945a, 1945b) under some simplifying assumptions. Recently, great strides have been made to expand and develop these ideas to solve many realistic problems about these stars.

In the present book, we shall attempt to record the basic mathematical methodologies, both analytical and numerical, developed during the second half of the present century for solving radiative transfer problems in spherically symmetric moving media. We shall consider only the effect of macroscopic velocity fields and omit discussions of turbulent motion. We believe that the works of the Nice and Heildelberg schools on turbulent motion in stellar envelopes merit treatment in a separate book.

In presenting this book, the authors recognise with great appreciation the monumental works of Mihalas (1978), Mihalas and Mihalas (1984) and Pomraning (1973) on the subject. We also record our appreciation to the lucid presentation of some of the numerical methods relevant to the solution of transfer problems in moving media contained in the two books edited by Kalkofen (1984, 1987). These books served us as valuable guiding references. They will continue to be fundamental references to all the contemporary workers on radiative transfer.

This book is addressed to graduate students and researchers in astrophysics, in particular to those interested in the study of radiative transfer in stellar atmospheres.

The book is divided into three parts. Part I contains the basic notions of radiation-matter interaction in participating media, the description of radiation field in moving media and the deduction of radiative transfer equations for spherically symmetric media in the presence of a velocity field. The transfer equations in lab frame, comoving frame and mixed frame are derived.

In Part II, we consider the basic mathematical methods for solving the transfer problems in extensive moving atmospheres when it is observed from the lab frame.

In Part III, we introduce the analytical and numerical methods for solving radiative transfer problems in spherically symmetric moving atmospheres, when the transfer equations are expressed in the comoving frame.

All attempts have been made to make the book self-contained. Cross references have been given wherever necessary. For the convenience of the readers, bibliographical references have also been given at the end of each chapter.

We are glad to record our grateful thanks to the referee for some helpful suggestions to improve the contents of the book.

We have great pleasure in acknowledging the immense help and encouragement we received from our wives and children during the preparation of this book. We also record our thanks and gratitude to our colleagues and friends for their assistance. Our special thanks goes to Dr. Chua Seng Kiat for his willing help in using the latex package and to Chan Lai Chee for helping us with the typing.

K.K. Sen
S.J. Wilson
December 1997.

References

Beals, C. (1929): MNRAS, **90**, 202.
Beals, C. (1930): P. Dom. Astrophys. Obs. Victoria, **4**, 271.
Beals, C. (1931): MNRAS, **91**, 966.
Beals, C. (1950): P. Dom. Astrophys. Obs. Victoria, **9**, 1.
Chandresekhar, S. (1934): MNRAS, **94**, 443.
Chandresekhar, S. (1945a): Rev. Mod. Phys., **17**, 138.
Chandresekhar, S. (1945b): Ap. J., **102**, 402.
Gerasimovič, B. (1934): Z. für. Astrophys., **7**, 335.
Kalkofen, W. (1984): Methods in Radiative Transfer, Cambridge University Press, Cambridge.
Kalkofen, W. (1987): Numerical Radiative Transfer, Cambridge University Press, Cambridge.
McCrea, W., Mitra, K. (1936): Z. für. Astrophys., **11**, 359.
Mihalas, D. (1978): Stellar Atmospheres (2nd edition). W.H. Freeman, San Francisco.
Mihalas, D., Mihalas, B.R.W. (1984): Foundations of Radiation Hydrodynamics, Oxford University Press, Oxford.
Pomraning, G. (1973): Radiation Hydrodynamics, Pergamen Press, Oxford.
Sobolev, V.V. (1960): Moving Envelopes of Stars, Harvard University Press, Cambridge.
Struve, O. (1929): Ap. J., **69**, 173.
Wilson, O. (1934): Ap. J., **80**, 250.

Table of Contents

Preface v
   References viii

1. **Basic Notions** 3
   1.1 Basic Notions of Radiative Transfer Problems in Stationary Media 3
   1.2 The Equation of Radiative Transfer 17
   1.3 Radiative Transfer in Stationary Spherical Medium 26
   1.4 Non-LTE Conditions and Statistical Equilibrium in Line Transfer Problems 32
   1.5 Outline of Methods of Solution of Radiative Transfer Equation in Spherically Symmetric Static Medium 40
   1.6 Basic Methods in Spherically Symmetric Static Media 41
   References 42

2. **Radiation Field and Transfer Equations in Moving Media** 43
   2.1 Radiative Transfer Equations in Lab Frame 44
   2.2 Transformation between Comoving and Lab frames . 50
   2.3 Transformation of Radiative Transfer Equation ... 56
   2.4 Mixed Equations 59
   2.5 Transfer Equations in CMF 64
   2.6 Moment Equations in Comoving Frame 69
   References 73

3. **Transfer Equations and Solutions in Lab Frame** .. 77
   3.1 The Escape Probability Method of Sobolev 84
   3.2 Generalised Sobolev Method 96
   3.3 Rosseland Cycle and Diffusion of Free Electrons ... 103
   References 105

4. **Impact Parameter Method for Transfer Equation in Lab Frame** 107
   4.1 Impact Parameter Method in Plane Geometry for Stationary Media 108
   4.2 Impact Parameter Method in Spherical Geometry for Stationary Media 118
   4.3 Impact Parameter Method for Spherically Symmetric Moving Media 126
   References 132

5. **Numerical Methods for Transfer Equations in Lab Frame** 133
   5.1 Direct Quadrature Method 133
   5.2 Escape Probability Method 142
   5.3 Core Saturation Method 144
   References 149

6. **Impact Parameter Method in CMF** 153
   6.1 Feautrier's Impact Parameter Method 154
   6.2 The Role of Aberration and Advection 164
   References 169

7. **Moment Equations in Comoving Frame** 171
   7.1 The Moment Equations and Their Solutions 172
   7.2 CMF Equations in Relativistic Flows 179
   7.3 Moment Methods Based on Generalised Eddington Approximation 186
   References 192

8. **Numerical Methods for Transfer Equations in CMF** 195
   8.1 Integral Operator Technique 195
   8.2 Adaptive Mesh Method 203
   8.3 The Method of Short Characteristic Rays 212
   8.4 Operator Perturbation Method 217
   References 244

Glossary of Physical Symbols 247

Index 253

PART

I

Chapter 1. Basic Notions

In this chapter, we gather together the physical notions for understanding the transport processes in a system involving radiation-matter interaction and state the mathematical tools for explaining them. The physical quantities occurring in subsequent chapters and those specifying the dynamical properties of the radiation field will be defined. The transfer equations will be formulated for the transport of energy through material media. The interaction between radiation and matter will be described by absorption, emission and scattering coefficients. Attention will be drawn to the equivalence of the radiative transfer equations to the Boltzmann equations of photon transport. The moment equations will be established. The known methods of solutions will be outlined. In this chapter, we shall assume the participating media to be stationary.

1.1 Basic Notions of Radiative Transfer Problems in Stationary Media

The Specific Intensity

We consider the one-dimensional structure of the radiation field and take note of the fact that the radiation field is a function of position and time. At a particular position and time, the radiation field is function of frequency and angle. It is in general expressed in terms of specific intensity $I(\mathbf{r}, t; \tilde{\mathbf{s}}, \nu)$ defined by

$$\frac{dE}{dt} = I(\mathbf{r}, t; \tilde{\mathbf{s}}, \nu)(\mathbf{n} \cdot \tilde{\mathbf{s}}) d\Sigma d\Omega d\nu; \qquad (1.1.1)$$

where $\frac{dE}{dt}$ is the rate of flow of radiation energy in the frequency interval $(\nu, \nu + d\nu)$ at a point **r** across an infinitesimal area $d\Sigma$ within an element of solid angle $d\Omega$. **n** is unit normal to $d\Sigma$ and $\tilde{\mathbf{s}}$, the unit vector in the direction of the pencil of radiation.

The frequency-integrated intensity $I(\mathbf{r}, t; \tilde{\mathbf{s}})$ is given by

$$I(\mathbf{r}, t; \tilde{\mathbf{s}}) = \int_0^\infty I(\mathbf{r}, t; \tilde{\mathbf{s}}, \nu) d\nu. \tag{1.1.2}$$

At a point, the radiation field is said to be isotropic, if the intensity is independent of direction.

The macroscopic description of the radiation field is given in terms of photon distribution function f_R occurring in Boltzmann equation. f_R is defined in such a way that the number of photons at a point **r** and time t in the frequency interval $(\nu, \nu + d\nu)$ contained per unit volume moving with velocity c in a direction $\tilde{\mathbf{s}}$ in a solid angle $d\Omega$ is $f_R(\mathbf{r}, t; \tilde{\mathbf{s}}, \nu)$.

Hence, the number of photons crossing an element of surface area $d\Sigma$ in time dt is equal to

$$(c\, dt)(\mathbf{n} \cdot \tilde{\mathbf{s}}) f_R d\Omega d\Sigma d\nu,$$

where **n** is the unit normal to the surface element $d\Sigma$.
Energy associated with each photon is $h\nu$. Then the energy dE transported across $d\Sigma$ is given by

$$\begin{aligned} dE &= f_R d\Omega d\nu d\Sigma (\mathbf{n} \cdot \tilde{\mathbf{s}})(c\, dt)(h\nu) \\ &= (\mathbf{n} \cdot \tilde{\mathbf{s}})(c\, h\nu) f_R d\Omega d\Sigma d\nu dt. \end{aligned} \tag{1.1.3}$$

From (1.1.3) and (1.1.1), we have

$$I(\mathbf{r}, t; \tilde{\mathbf{s}}, \nu) = (c\, h\nu) \cdot f_R(\mathbf{r}, t; \tilde{\mathbf{s}}, \nu), \tag{1.1.4}$$

where $f_R(\mathbf{r}, t; \tilde{\mathbf{s}}, \nu)$ is called the relativistically invariant photon distribution function.

In plane geometry for one dimensional case, assuming homogeneity of material in the horizontal plane (x, y), the radiation field will have azimuthal symmetry about **k**, the unit vector along the Z-direction The specific intensity in this case can be written as

$$I = I(z, t; \mu, \nu) \tag{1.1.5}$$

1.1 Basic Notions of Radiative Transfer Problems in Stationary Media

where
$$\mu = \tilde{\mathbf{s}} \cdot \mathbf{k} = \cos\theta. \tag{1.1.6}$$

In spherical geometry, the transport of photons is expressed in polar coordinates (r, θ, φ), where r denotes the position and (θ, φ) denotes the direction of the ray. θ is the polar angle and φ the azimuth angle. In the case of spherically symmetric radiation field the specific intensity is azimuth independent and we can write it as

$$I = I(r, t; \mu, \nu), \tag{1.1.7}$$

where $\mu = \tilde{\mathbf{s}} \cdot \tilde{\mathbf{r}}$, $\tilde{\mathbf{r}}$ being the radial unit vector.

The Mean Intensity, Energy Density, Flux, Momentum density and Pressure tensor of Radiation

The Mean Intensity

The mean intensity is the average of specific intensity integrated over all solid angles. Hence, the mean intensity $J(\mathbf{r}, t; \nu)$ is given by

$$J(\mathbf{r}, t; \nu) = \frac{1}{4\pi} \oint I(\mathbf{r}, t : \tilde{\mathbf{s}}, \nu) d\Omega. \tag{1.1.8}$$

It is the zeroth angular moment of specific intensity. The frequency-integrated mean intensity reads as

$$\tilde{J}(\mathbf{r}, t) = \int_0^\infty J(\mathbf{r}, t, \nu) d\nu. \tag{1.1.9}$$

In plane parallel media, the radiation field is symmetric about Z-axis and the mean intensity can be written as

$$J(z, t; \nu) = \oint I(z, t; \tilde{\mathbf{s}}, \nu) d\Omega$$
$$= \frac{1}{4\pi} \int_0^{2\pi} d\varphi \int_{-1}^1 I(z, t; \mu, \nu) d\mu \tag{1.1.10}$$

i.e.
$$J(z, t; \nu) = \frac{1}{2} \int_{-1}^1 I(z, t; \mu, \nu) d\mu, \tag{1.1.11}$$

where $d\Omega = sin\theta d\theta d\varphi = -d\mu d\varphi$. The frequency-integrated intensity for the plane parallel case is

$$\tilde{J}(z,t) = \int_0^\infty J(z,t,\nu)d\nu. \qquad (1.1.12)$$

In spherical medium with spherically symmetric radiation field, the mean intensity can be written as

$$J(\mathbf{r},t;\nu) = \frac{1}{2}\int_{-1}^1 I(\mathbf{r},t;\mu,\nu)d\mu, \qquad (1.1.13)$$

and the frequency-integrated mean intensity is given by

$$\tilde{J}(\mathbf{r},t) = \int_0^\infty J(\mathbf{r},t,\nu)d\nu. \qquad (1.1.14)$$

Energy Density of Radiation

Remembering that the number of photons per unit volume at (r,t) within frequency interval $(\nu, \nu+d\nu)$ within solid angle $d\Omega$ around $\tilde{\mathbf{s}}$ is given by the photon distribution function $f_R(\mathbf{r},t;\tilde{\mathbf{s}},\nu)d\Omega d\nu$, the monochromatic radiant energy $E(\mathbf{r},t;\nu)$ at ν, is given by

$$E_\nu = E(\mathbf{r},t;\nu) = (h\nu)\oint f_R(\mathbf{r},t;\tilde{\mathbf{s}},\nu)d\Omega. \qquad (1.1.15)$$

Then from (1.1.4) and (1.1.8),

$$\begin{aligned} E(\mathbf{r},t;\nu) &= \frac{1}{c}\oint I(\mathbf{r},t;\mu,\nu)d\Omega \\ &= (\frac{4\pi}{c})J(\mathbf{r},t;\nu). \end{aligned} \qquad (1.1.16)$$

Hence, the frequency-integrated total energy density E is given by

$$\begin{aligned} E &= \int_0^\infty E(\mathbf{r},t;\nu)d\nu \\ &= (\frac{4\pi}{c})\int_0^\infty J(\mathbf{r},t;\nu)d\nu \\ &= (\frac{4\pi}{c})\tilde{J}(\mathbf{r},t). \end{aligned} \qquad (1.1.17)$$

Radiation Flux, Momentum Density and Pressure tensor

The rate of flow of radiant energy in the frequency interval $(\nu, \nu + d\nu)$ across an elemental area $d\Sigma$ is equal to

$$(\oint I(\mathbf{r},t;\tilde{\mathbf{s}},\nu)(\mathbf{n}\cdot\tilde{\mathbf{s}})d\Omega)d\Sigma d\nu = (\tilde{F}_\nu)d\Sigma d\nu \qquad (1.1.18)$$

where $\tilde{F}_\nu$, the monochromatic flux is given by

$$\tilde{F}_\nu = \oint I(\mathbf{r},t;\tilde{\mathbf{s}},\nu)(\mathbf{n}\cdot\tilde{\mathbf{s}})d\Omega. \qquad (1.1.19)$$

It represents the rate of flow of radiant energy per unit area across $d\Sigma$ per unit frequency interval at frequency ν. Now remembering that $d\Omega = \sin\theta d\theta d\varphi$, where θ and φ are the polar and azimuthal angles indicating the direction of the radiation field measured relative to the radial unit vector, we have

$$\tilde{F}_\nu = \int_0^\pi d\theta \int_0^{2\pi} d\varphi\, I(r,t;\theta,\varphi,\nu)\cos\theta\sin\varphi. \qquad (1.1.20)$$

For axially symmetric radiation field

$$\tilde{F}_\nu = \tilde{F}_\nu(r,t;\nu) = (2\pi)\int_{-1}^{1} I(r,t;\mu,\nu)\mu d\mu. \qquad (1.1.21)$$

In the astrophysical context, it is convenient to write $\tilde{F}_\nu = \pi F_\nu$ where

$$F_\nu = 2\int_{-1}^{1} I(r,t;\mu,\nu)\mu d\mu. \qquad (1.1.22)$$

Furthermore, Eddington introduced the quantity $H_\nu = (1/4)F_\nu$ given by

$$H_\nu = \frac{1}{2}\int_{-1}^{1} I(r,t;\mu,\nu)\mu d\mu, \qquad (1.1.23)$$

to have similarity between the definitions of mean intensity J_ν and H_ν. The frequency-integrated radiation flux is given by

$$\tilde{F} = \int_0^\infty \tilde{F}_\nu\, d\nu, \qquad F = \int_0^\infty F_\nu\, d\nu \qquad (1.1.24)$$

$$H = \int_0^\infty H_\nu\, d\nu. \qquad (1.1.25)$$

Radiation Pressure Tensor or Radiation Stress Tensor

The concept of radiation pressure tensor $\mathcal{P}$ is associated with the idea of transport of momentum. $\mathcal{P}$ is a tensor whose specific component P^{ij} gives the net rate of transfer of ith component of momentum per unit area of a surface placed normal to the jth coordinate axis.

$$P^{ij}(\mathbf{r},t;\nu) = \frac{1}{c}\oint I(\mathbf{r},t;\tilde{\mathbf{s}},\nu)\tilde{s}^i\tilde{s}^j d\Omega, \qquad (1.1.26)$$

or writing in a dyadic form, we have

$$\mathcal{P}_\nu = \mathcal{P}(\mathbf{r},t;\nu) = \frac{1}{c}\oint I(\mathbf{r},t;\tilde{\mathbf{s}},\nu)\tilde{\mathbf{s}}\tilde{\mathbf{s}} d\Omega. \qquad (1.1.27)$$

It is seen that $\mathcal{P}$ is a symmetric quantity implying that $P^{ij} = P^{ji}$. It is connected with the second angular moment of specific intensity. In one dimensional case $\mathcal{P}$ can be considered as a scalar and can be expressed as

$$P_\nu = (\frac{4\pi}{c})K_\nu \qquad (1.1.28)$$

where

$$K_\nu = \frac{1}{2}\int_{-1}^{1} I(r,t;\mu,\nu)\mu^2 d\mu. \qquad (1.1.29)$$

[cf. Mihalas and Mihalas (1984) p. 314–318].

The radiation pressure tensor is not always isotropic. However, in the case of isotropic radiation field, Eddinton approximation $3K_\nu = J_\nu$ holds good. Then from (1.1.16), (1.1.17) and (1.1.28) we have for the Eddington factor f_ν and P_ν

$$f_\nu = \frac{K(r,t;,\nu)}{J(r,t;\nu)} = \frac{1}{3}, \qquad (1.1.30)$$

$$P_\nu = (\frac{1}{3})E_\nu. \qquad (1.1.31)$$

In general, f_ν deviates from $(1/3)$ and the amount of deviation gives the measure of the anisotropy of the radiation field. f_ν tends to the value $(1/3)$ in the far interior of the stellar atmosphere while near the upper bounding surface f_ν is larger than $(1/3)$. f_ν lies between $(1/3)$ and unity.

1.1 Basic Notions of Radiative Transfer Problems in Stationary Media

The total radiation pressure, in the one-dimensional case is obtained by integrating $P(r,t;\nu)$ over frequency ν and can be expressed as a scalar.

$$P = P(r,t) = \int_0^\infty P(r,t;\nu)d\nu = \frac{4\pi}{c}\int K(r,t;\nu)d\nu. \quad (1.1.32)$$

The total radiation pressure tensor or radiation stress tensor is given by

$$\mathcal{P} = \mathcal{P}(r,t) = \frac{1}{c}\int_0^\infty d\nu \oint d\Omega I(\mathbf{r},t;\tilde{\mathbf{s}},\nu)\tilde{\mathbf{s}}\tilde{\mathbf{s}}. \quad (1.1.33)$$

Radiation–Matter Interaction

In the study of radiative transfer in stellar atmospheres, the interaction of radiation with matter is described by macroscopic processes of absorption, emission and scattering.

Absorption

A pencil of radiation loses a portion of its energy during its passage through matter and does not always reappear in the same frequency as a scattered ray. This loss is described as "true" absorption in the sense that it is transformed into other forms of energy. The measure of this loss is done through the concept of "absorption coefficient" which is defined below. It is the fraction of the total loss of energy from the pencil of radiation. The loss, in addition, can also be due to scattering of radiation. The loss of energy of radiation from the beam due to "true" or "thermal" absorption per unit volume of the medium, per unit time, unit solid angle and unit frequency interval is

$$k(\mathbf{r},t;\tilde{\mathbf{s}},\nu)I(\mathbf{r},t;\tilde{\mathbf{s}},\nu), \quad (1.1.34)$$

where $k(\mathbf{r},t;\tilde{\mathbf{s}},\nu)$ is the absorption coefficient. The mass absorption coefficient is $k\rho$, ρ being the density of the medium. The absorption coefficient $k(\mathbf{r},t;\tilde{\mathbf{s}},\nu)$ is isotropic in a homogeneous fluid medium at rest, in the lab frame. However, if the medium is in motion with respect to the observer, the Doppler shift of frequencies has to be taken into account.

A photon with frequency ν moving in the direction $\tilde{\mathbf{s}}$ will appear to have a frequency ν_0 in the fluid frame moving with velocity

V with respect to the lab frame. From the Doppler effect in the classical sense, ν and ν_0 are connected as

$$\nu_0 = \nu(1 - \tilde{\mathbf{s}} \cdot \frac{\mathbf{V}}{c}). \tag{1.1.35}$$

The absorption coefficient in this case becomes angle dependent and hence anisotropic in the observer's frame. For complete description of the radiation field in the moving media, the effects of "aberration" and "advection" have also to be taken into account. Both of these factors contribute to the anisotropy of the absorption coefficient.

Emission

The generation of energy in the bulk of a material medium, "thermal" or otherwise, may stimulate excitation and de-excitation of atoms constituting it and add to the pencil of radiant energy. This process of gain of energy in a beam is called "true" or "thermal" emission. This does not include the contribution to the pencil of radiation from the rays in other directions by way of scattering. The emission coefficient or emissivity $\eta^T(\mathbf{r}, t; \tilde{\mathbf{s}}, \nu)$ is defined such that the amount of radiant energy dE emitted by matter of an elementary length ds, cross section $d\Sigma$ into a solid angle $d\Omega$ around $\tilde{\mathbf{s}}$ in the frequency interval $(\nu, \nu + d\nu)$ in time dt is

$$dE = \eta^T(\mathbf{r}, t; \tilde{\mathbf{s}}, \nu) ds d\Sigma d\Omega d\nu dt. \tag{1.1.36}$$

The emissivity is isotropic in a homogeneous fluid medium at rest. However, if the medium is in motion with respect to the observer, the emissivity $\eta^T(\mathbf{r}, t; \tilde{\mathbf{s}}, \nu)$ becomes angle dependent and anisotropic due to Doppler shift, aberration and advection.

Redistribution Function, Phase Function and Scattering

The scattering process can be described in terms of a "redistribution function". The scattering process is coherent when it involves only the change of direction of a pencil and not its frequency. When both direction and frequency change the scattering process is said to be non-coherent. Let $R(\tilde{\mathbf{s}}', \nu'; \tilde{\mathbf{s}}, \nu)$ denote the redistribution function. Then the probability that a photon in the frequency interval $(\nu', \nu' + d\nu')$ in the direction $\tilde{\mathbf{s}}'$ within the solid angle $d\Omega'$

1.1 Basic Notions of Radiative Transfer Problems in Stationary Media

will be scattered in the direction $\tilde{s}$ within the solid angle $d\Omega$ in the frequency interval $(\nu, \nu + d\nu)$ is given by

$$R(\tilde{s}', \nu'; \tilde{s}, \nu)(\frac{d\Omega'}{4\pi})(\frac{d\Omega}{4\pi}).$$

$R(\tilde{s}', \nu'; \tilde{s}, \nu)$ is normalised such that

$$(\frac{1}{4\pi})^2 \oint d\Omega' \oint d\Omega \int_0^\infty d\nu' \int_0^\infty R(\tilde{s}', \nu'; \tilde{s}, \nu) d\nu = 1. \quad (1.1.37)$$

For the process of scattering, the normalised absorption and emission profile can be expressed in terms of the redistribution function $R(\tilde{s}', \nu'; \tilde{s}, \nu)$.

Let $\phi(\nu')$ be the normalised absorption profile. Then

$$\phi(\nu') = \frac{1}{4\pi} \oint d\Omega \int_0^\infty R(\tilde{s}', \nu'; \tilde{s}, \nu) d\nu, \quad (1.1.38)$$

where

$$\int_0^\infty \phi(\nu') d\nu' = 1. \quad (1.1.39)$$

We now introduce the scattering coefficient $\sigma(\mathbf{r}, t; \tilde{s}, \nu)$. From a beam of specific intensity $I(\mathbf{r}, t; \tilde{s}', \nu')$, the amount of energy removed at frequency interval $(\nu', \nu' + d\nu')$ in solid angle $d\Omega'$ and scattered into a solid angle $d\Omega$ in the frequency interval $(\nu, \nu + d\nu)$ is given by

$$\sigma(\mathbf{r}, t; \tilde{s}, \nu) \, R(\tilde{s}', \nu'; \tilde{s}, \nu) \, I(\mathbf{r}, t; \tilde{s}', \nu') \, d\nu' d\nu (\frac{d\Omega'}{4\pi})(\frac{d\Omega}{4\pi}). \quad (1.1.40)$$

Integrating over all the incoming angles and frequencies which are denoted by primes in (1.1.40), we obtain a measure of the radiant energy added to a pencil of radiation $I(\mathbf{r}, t; \tilde{s}.\nu)$ from the beam in other frequencies and directions by way of scattering. Then $\eta^s(\mathbf{r}, t; \tilde{s}, \nu)$, the scattering emissivity or emission coefficient of scattering can be defined in general form as

$$\eta^s(\mathbf{r}, t; \tilde{s}, \nu) d\nu (\frac{d\Omega}{4\pi})$$
$$= \sigma(\mathbf{r}, t; \tilde{s}, \nu) d\nu (\frac{d\Omega}{4\pi}) \oint (\frac{d\Omega'}{4\pi}) \int_0^\infty R(\tilde{s}', \nu'; \tilde{s}, \nu) I(\mathbf{r}, t; \tilde{s}', \nu') d\nu'. \quad (1.1.41)$$

Many problems may require the use of η^s much more restricted than defined in (1.1.41). For example, in some problems, one may be interested only in redistribution of frequency or in coherent distribution of angles. We take note of the fact that the redistribution function depends only on the angle between **s** and **s'** and define the angle averaged time independent redistribution function $R(\nu', \nu)$ as

$$\begin{aligned} R(\nu', \nu) &= \frac{1}{4\pi} \oint R(\tilde{\mathbf{s}}', \nu'; \tilde{\mathbf{s}}, \nu) d\Omega' \\ &= \frac{1}{4\pi} \oint R(\tilde{\mathbf{s}}, \nu; \tilde{\mathbf{s}}', \nu') d\Omega \\ &= R(\nu, \nu'), \end{aligned} \qquad (1.1.42)$$

implying the reciprocity of the redistribution function, i.e.

$$R(\nu', \nu) = R(\nu, \nu').$$

Then we may write

$$\eta^s(\mathbf{r}, \nu) = \sigma(\mathbf{r}) \int_0^\infty R(\nu', \nu) J(\mathbf{r}, \nu') d\nu', \qquad (1.1.43)$$

where the mean intensity

$$J(\mathbf{r}, \nu') = \oint I(\mathbf{r}, \tilde{\mathbf{s}}, \nu') \left(\frac{d\Omega'}{4\pi}\right). \qquad (1.1.44)$$

From (1.1.42), we may conclude that $R(\nu', \nu)$ is the measure of the probability of a photon in the frequency range $(\nu', \nu' + d\nu')$ being redistributed to the frequency range $(\nu, \nu + d\nu)$. The term "probability" is justified as it satisfies the following normality condition [cf. (1.1.37)]

$$\int_0^\infty d\nu' \int_0^\infty R(\nu', \nu) d\nu = \int_0^\infty \phi(\nu') d\nu' = 1, \qquad (1.1.45)$$

where

$$\phi(\nu') = \int_0^\infty R(\nu', \nu) d\nu. \qquad (1.1.46)$$

Let $\psi(\nu)$ be the angle averaged emission profile function. It is given by

$$\psi(\nu) = \frac{\eta^s(\mathbf{r},\nu)}{\int_0^\infty \eta^s(\mathbf{r},\nu)d\nu}$$
$$= \frac{\int_0^\infty R(\nu',\nu)J(\mathbf{r},\nu')d\nu'}{\int_0^\infty \phi(\nu')J(\mathbf{r},\nu')d\nu'}. \quad (1.1.47)$$

Equation (1.1.47) suggests the dependence of emission profile on the profile of the incident radiation.

In the case of frequency independent radiation intensity, since

$$\psi(\nu) = \int_0^\infty R(\nu',\nu)d\nu' \quad (1.1.48)$$

$$\psi(\nu) = \psi(\nu'). \quad (1.1.49)$$

These conditions hold in the case of thermodynamic equilibrium. In the special case of intensity being frequency independent, $\psi(\nu) = \phi(\nu)$. For coherent scattering, $\nu' = \nu$, and

$$R(\tilde{\mathbf{s}}',\nu';\tilde{\mathbf{s}},\nu) = p(\tilde{\mathbf{s}}',\tilde{\mathbf{s}})\phi(\nu')\delta(\nu - \nu'), \quad (1.1.50)$$

where $p(\tilde{\mathbf{s}}',\tilde{\mathbf{s}})$ is the phase function and $\delta(\nu - \nu')$ is the Kronecker delta function. $p(\tilde{\mathbf{s}}',\tilde{\mathbf{s}})$ is normalised as

$$\frac{1}{4\pi}\oint p(\tilde{\mathbf{s}}',\tilde{\mathbf{s}})d\Omega' = 1. \quad (1.1.51)$$

For isotropic scattering, $p(\tilde{\mathbf{s}}',\tilde{\mathbf{s}}) = 1.$ and

$$R(\tilde{\mathbf{s}}',\nu';\tilde{\mathbf{s}},\nu) = R(\nu',\nu)$$
$$= \phi(\nu')\delta(\nu - \nu'). \quad (1.1.52)$$

Incidentally it may be mentioned that when the redistribution function involves distribution of both angle and frequency, the solution of radiative transfer equation becomes extremely complex [cf. Mihalas (1978) p. 412–433]. It becomes particularly so in the case of Doppler-shifted redistribution in moving atmospheres when studied in lab frame, where emission and absorption coefficients become angle dependent. In the study of the transfer problems in spherically symmetric moving media, we shall confine ourselves to complete angle averaged redistribution functions. However, it should be borne in mind that the condition of equality of absorption and emission profiles $\phi(\nu) = \psi(\nu)$ seldom holds good strictly in real situations and complete redistribution in co-moving frame hardly ensures complete redistribution in lab frame. However in most cases, complete redistribution assumption in lab frame leads to fairly good approximation and can be written from (1.1.37), (1.1.38) and (1.1.42) as

$$R(\nu', \nu) = \phi(\nu)\phi(\nu').$$

Attenuation or Extinction Coefficient (or Opacity) and Total Emissivity

The total loss of intensity from a pencil of radiation due to "thermal" absorption and scattering is usually expressed in terms of an attenuation coefficient, (extinction coefficient or opacity) $\chi(\mathbf{r}, t; \tilde{\mathbf{s}}, \nu)$ given by

$$\chi(\mathbf{r}, t; \tilde{\mathbf{s}}, \nu) = k(\mathbf{r}, t; \tilde{\mathbf{s}}, \nu) + \sigma(\mathbf{r}, t; \tilde{\mathbf{s}}, \nu), \tag{1.1.53}$$

where k is the thermal absorption coefficient and σ is the scattering coefficient. The mean free path of the photon is given by $\lambda = \frac{1}{\chi}$. Similarly, the total emissivity can be expressed as the sum of the thermal and scattering emissivities, η^T and η^s, respectively. Thus, total emissivity η is given by

$$\eta(\mathbf{r}, t; \tilde{\mathbf{s}}, \nu) = \eta^T(\mathbf{r}, t; \tilde{\mathbf{s}}, \nu) + \eta^s(\mathbf{r}, t; \tilde{\mathbf{s}}, \nu). \tag{1.1.54}$$

If the scattering process is conservative, the amount of energy from the beam is re-emitted immediately. The re-emission of total energy attenuated from the beam is described by the redistribution function $R(\tilde{\mathbf{s}}', \nu'; \tilde{\mathbf{s}}, \nu)$. This gives the measure of the probability that a photon $(\tilde{\mathbf{s}}', \nu')$ absorbed, reappears as a photon $(\tilde{\mathbf{s}}, \nu)$.

The redistribution can be partial or complete. The redistribution is complete when the photons are isotropically emitted and randomly redistributed over the line profile. In many cases of physical interest, the complete redistribution of photons appears to be a valid assumption. This condition obtains when within the Doppler core of line, the atoms excited by absorption of photons undergo a number of collisions before re-emission of any photons.

1.1 Basic Notions of Radiative Transfer Problems in Stationary Media

The collision processes practically destroys any correlation between the frequencies of absorbed and emitted photons. It should also be emphasised that scattering is essentially non-local process depending on the radiation intensity $I(\mathbf{r}, t; , \tilde{\mathbf{s}}, \nu)$ at any point $\mathbf{r}$ contributed from other points. During this interaction between the radiation at two points, the thermal radiation at that point practically plays no role. In general, both in the case of line and continuum radiation, thermal and scattered radiation make partial contributions.

Line Profile in Moving Medium

For a medium with velocity $\mathbf{V}(\mathbf{r})$ relative to an external observer at rest (lab frame), photon frequencies are Doppler shifted between the lab frame and frame of the atoms constituting the moving media. Let ν be the frequency in the lab frame. Then in the atom's frame (a frame comoving with the atom) the frequency ν' at which a photon traveling in a direction $\mathbf{n}$ was emitted or can be absorbed is given by [cf. (1.1.35)]

$$\nu' = \nu - \nu_0(\frac{\mathbf{n} \cdot \mathbf{V}}{c}), \tag{1.1.55}$$

where ν_0 is the line centre frequency. It is customary to measure the frequency displacement from the line centre in units of Doppler width $\Delta\nu_D = (\nu_0 v_{th})/c$ where v_{th} is a thermal velocity parameter. Velocity is also measured in the same units i.e. $V = v/v_{th}$. The transformation between comoving frame and lab frame frequencies is given by

$$x' = x - \mu V, \tag{1.1.56}$$

where

$$x = \frac{(\nu - \nu_0)}{\Delta\nu_D}.$$

The normalised line profile is defined as

$$\phi(\mathbf{r}, x') = \phi(\mathbf{r}, x - \mu V). \tag{1.1.57}$$

where V is the velocity of the comoving frame. The transformation of material and radiation variables from lab frame to comoving frame will be taken up in chapter 2 [cf. section 2.2] and will be used in subsequent chapters.

Source Function in Time Independent Radiative Transfer in Stationary Medium

In stationary medium opacity and emissivity are angle independent. Hence the source function $S(\mathbf{r}, \nu)$ at a particular point $\mathbf{r}$ at frequency ν, which is the ratio of total emissivity $\eta(\mathbf{r}, \nu)$ to opacity $\chi(\mathbf{r}, \nu)$, is given by

$$S(\mathbf{r}, \nu) = \frac{\eta(\mathbf{r}, \nu)}{\chi(\mathbf{r}, \nu)}. \qquad (1.1.58)$$

If further, the medium is in local thermodynamic equilibrium (LTE), a local temperature $T(\mathbf{r})$ can be assigned at each point $\mathbf{r}$ of the medium. Under these conditions, Kirchoff's law is valid and $S(\mathbf{r}, \nu)$ can be written as

$$\begin{aligned} S(\mathbf{r}, \nu) &= \frac{\eta(\mathbf{r}, \nu)}{\chi(\mathbf{r}, \nu)} \\ &= B_\nu(T), \end{aligned} \qquad (1.1.59)$$

where $B_\nu(T)$ is the Planck function. Here,

$$B_\nu(T) = (\frac{2h\nu^3}{c^2})[\exp{(h\nu/kT)} - 1]^{-1}, \qquad (1.1.60)$$

where h is the Planck constant and k is the Boltzmann constant. For scattering medium involving coherent scattering ($\nu' = \nu$) and no thermal emission, the source function

$$S(\mathbf{r}, \nu) = \omega(\mathbf{r}, \nu) \oint p(\tilde{\mathbf{s}}', \tilde{\mathbf{s}}) I(\mathbf{r}, \tilde{\mathbf{s}}', \nu) \frac{d\Omega'}{4\pi}. \qquad (1.1.61)$$

In (1.1.61), emission and absorption profiles have been taken to be the same. $\omega(\mathbf{r}, \nu)$ is the albedo for single scattering and $p(\tilde{\mathbf{s}}', \tilde{\mathbf{s}})$ is the phase function defined in (1.1.50). Here

$$\omega(\mathbf{r}, \nu) = \frac{\sigma(\mathbf{r}, \nu)}{\chi(\mathbf{r}, \nu)}.$$

$\omega(\mathbf{r}, \nu) = 1$ represents perfect scattering and $\omega(\mathbf{r}, \nu) = 0$ represents medium involving thermal absorption and emission.

1.2 The Equation of Radiative Transfer

The equation of radiative transfer is the fundamental equation for studying the change of specific intensity of radiation from point to point in participating medium. The interaction of radiation and matter is characterised by absorption, emission and scattering. In what follows, we shall write down the radiative transfer equations in stationary media, and put it in forms appropriate for plane parallel and spherically symmetric media. Attention will also be drawn to well known approximations introduced to the general form of transfer equations in both plane parallel and spherical geometry.

We consider the passage of a pencil of radiation through a cylindrical element of material of volume $d\Sigma ds$ at a point $\mathbf{r}$ in a direction $\tilde{\mathbf{s}}$. $d\Sigma$ is an infinitesimal area normal to the direction of $\tilde{\mathbf{s}}$. ds is the elementary length of the cylinder.

Let $I(\mathbf{r};\tilde{\mathbf{s}},\nu)$ be the specific intensity of the radiation pencil of frequency ν incident at $\mathbf{r}$ in a direction $\tilde{\mathbf{s}}$. The rate at which the radiant energy is attenuated in passing through the volume $d\Sigma ds$ is given by

$$\chi(\mathbf{r},\nu)ds[I(\mathbf{r};\tilde{\mathbf{s}},\nu)d\Sigma d\Omega d\nu].$$

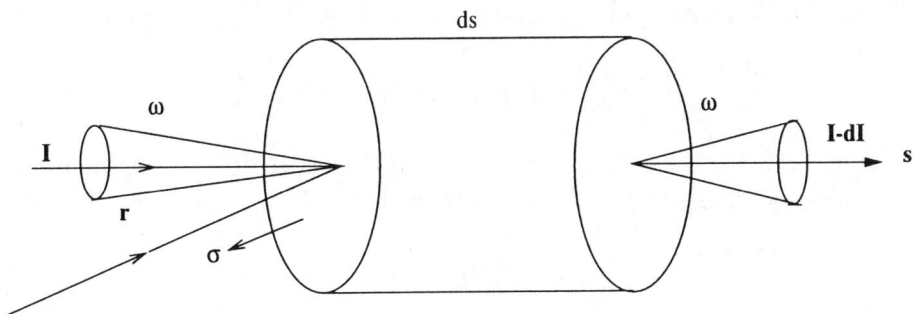

Figure 1.1. Schematic diagram for the propagation of radiative intensity

We recall that $I(\mathbf{r}; \tilde{\mathbf{s}}, \nu)$ represents the rate of flow of energy of frequency ν per unit frequency interval within unit solid angle across unit area normal to the direction $\tilde{\mathbf{s}}$ at a point $\mathbf{r}$.

The attenuated radiation may be re-emitted as a thermal radiation and/or scattered radiation from other directions. The process of scattering may be coherent or non-coherent. The measure of contribution of scattering to the total emission is done through an albedo for single scattering $\omega(\mathbf{r})$ which is the ratio of scattering coefficient $\sigma(\mathbf{r})$ to attenuation coefficient $\chi(\mathbf{r})$. In fact, the process of scattering contributes both to the emission and attenuation of radiation and these contributions can be expressed in terms of the redistribution functions or phase functions.

The rate at which radiant energy is emitted within the volume element $d\Sigma ds$ in solid angle $d\Omega$ in the frequency interval $d\nu$ is

$$\eta(\mathbf{r}, \nu) d\Sigma ds d\Omega d\nu,$$

where $\eta(\mathbf{r}, \nu)$ is the total emission coefficient. Hence the balance equation after the pencil has passed through the elementary cylinder $d\Sigma ds$ is given by

$$\begin{aligned} dI(\mathbf{r}, \tilde{\mathbf{s}}, \nu) d\Sigma d\Omega d\nu &= -\chi(\mathbf{r}, \nu) ds I(\mathbf{r}, \tilde{\mathbf{s}}, \nu) d\Sigma d\Omega d\nu \\ &\quad + \eta(\mathbf{r}, \nu) ds d\Sigma d\Omega d\nu, \end{aligned} \quad (1.2.1)$$

where $dI(\mathbf{r}, \tilde{\mathbf{s}}, \nu)$ is the change in specific intensity in passage through the volume element of the medium.

In the limit as $ds \to 0$, we have

$$\frac{dI(\mathbf{r}, \tilde{\mathbf{s}}, \nu)}{ds} = -\chi(\mathbf{r}, \nu) I(\mathbf{r}, \tilde{\mathbf{s}}, \nu) + \eta(\mathbf{r}, \nu). \quad (1.2.2)$$

Using (1.1.53) and (1.1.54), the radiative transfer equation in a stationary medium can be written in the following general form.

$$\frac{dI(\mathbf{r}, \tilde{\mathbf{s}}, \nu)}{ds} = -\chi(\mathbf{r}, \nu)[I(\mathbf{r}, \tilde{\mathbf{s}}, \nu) - S(\mathbf{r}, \nu)], \quad (1.2.3)$$

where the source function

$$S(\mathbf{r}, \nu) = \frac{\eta(\mathbf{r}, \nu)}{\chi(\mathbf{r}, \nu)} \quad (1.2.4)$$

1.2 The Equation of Radiative Transfer

with
$$\chi(\mathbf{r},\nu) = k(\mathbf{r},\nu) + \sigma(\mathbf{r},\nu)$$
$$\eta(\mathbf{r},\nu) = \eta^T(\mathbf{r},\nu) + \eta^s(\mathbf{r},\nu). \qquad (1.2.5)$$

$k(\mathbf{r},\nu)$, $\sigma(\mathbf{r},\nu)$, $\eta^T(\mathbf{r},\nu)$, and $\eta^s(\mathbf{r},\nu)$ are defined in (1.1.34), (1.1.40), (1.1.54) and (1.1.41). Equation (1.2.3) is the general form of transfer equation when the specific intensity $I(\mathbf{r},\tilde{\mathbf{s}},\nu)$ is time independent.

Some Special Cases

(i) In local thermodynamic equilibrium (LTE), when the radiation-matter interaction involves only thermal absorption and emission, Kirchoff's law holds good and the source function $S(\mathbf{r},\nu)$ is given by
$$S(\mathbf{r},\nu) = B_\nu(T), \qquad (1.2.6)$$
where the Planck function, $B_\nu(T)$ is defined in (1.1.60).

The equation of transfer in this case reads as
$$\frac{dI((\mathbf{r},\tilde{\mathbf{s}},\nu)}{ds} = -\chi(\mathbf{r},\nu)[I(\mathbf{r},\tilde{\mathbf{s}},\nu) - B_\nu(T)]. \qquad (1.2.7)$$

(ii) A purely scattering medium implies $\chi(\mathbf{r},\nu) = \sigma(\mathbf{r},\nu)$ and $\omega(\mathbf{r}) = 1$. The transfer equation in the purely scattering case is
$$\frac{1}{\chi(\mathbf{r},\nu)}[\frac{dI((\mathbf{r},\tilde{\mathbf{s}},\nu)}{ds}] = -I(\mathbf{r},\tilde{\mathbf{s}},\nu) + \frac{1}{4\pi}\oint p(\tilde{\mathbf{s}}',\tilde{\mathbf{s}})I(\mathbf{r},\tilde{\mathbf{s}}',\nu)d\Omega'. \qquad (1.2.8)$$

(iii) When the radiation-matter interaction involves both the thermal and scattering processes, partial contribution of both of them is taken into account in writing the radiative transfer equation. Assuming the medium to be in local thermodynamic equilibrium and taking the scattering process as coherent, the transfer equation takes the form

$$\frac{1}{\chi(\mathbf{r},\nu)}\left[\frac{dI(\mathbf{r},\tilde{\mathbf{s}},\nu)}{ds}\right] = -I(\mathbf{r},\tilde{\mathbf{s}},\nu) + S(\mathbf{r},\nu), \qquad (1.2.9)$$

where

$$S(\mathbf{r},\nu) = [1 - \omega(\mathbf{r},\nu)]B_\nu(\mathbf{r}) + \frac{\omega(\mathbf{r},\nu)}{4\pi} \oint p(\tilde{\mathbf{s}}',\tilde{\mathbf{s}})I(\mathbf{r},\tilde{\mathbf{s}}',\nu)d\Omega'. \qquad (1.2.10)$$

We recall that

$$[1 - \omega(\mathbf{r},\nu)] = \frac{k(\mathbf{r},\nu)}{\chi(\mathbf{r},\nu)}.$$

(iv) If in addition, the contribution to source function $S(\mathbf{r},\nu)$ comes also from external radiation incident on the outer bounding surface, the source function then can be written as

$$S(\mathbf{r},\nu) = \frac{\omega(\mathbf{r},\nu)}{4\pi} \oint p(\tilde{\mathbf{s}}',\tilde{\mathbf{s}})I(\mathbf{r},\tilde{\mathbf{s}},\nu)d\Omega' + B(\mathbf{r},\nu), \qquad (1.2.11)$$

where

$$B(\mathbf{r},\nu) = B_1(\mathbf{r},\nu) + B_0(\mathbf{r},\nu). \qquad (1.2.12)$$

$B_0(\mathbf{r},\nu)$ is the contribution from the reduced incident radiation from the external sources at $\mathbf{r}$ and frequency ν and $B_1(\mathbf{r},\nu)$ is that from internal sources other than scattering. $B_0(\mathbf{r},\nu)$ and $B_1(\mathbf{r},\nu)$ depend on the physical nature of the problem at hand and are specified.

Grey Medium Approximation

In radiative transfer in stellar atmospheres and in heat transfer, there are many problems in which radiative properties of the medium can be taken to be frequency independent. Such media are termed as "grey" media. In grey approximation, it is customary to use frequency integrated radiative and material properties in the transfer equation. For example, we may write

$$I(\mathbf{r},\tilde{\mathbf{s}}) = \int_0^\infty I(\mathbf{r},\tilde{\mathbf{s}},\nu)d\nu. \qquad (1.2.13)$$

In particular, in heat transfer problems, we can write for the thermal part of the source function the frequency-integrated Planck function as

$$\int_0^\infty B_\nu(T)d\nu = \frac{n^2\bar{\sigma}T^4}{\pi}, \qquad (1.2.14)$$

where n is the refractive index of the medium and $\bar{\sigma}$ the Stephan constant.

The radiative transfer equation in this case, can be written as

$$\frac{1}{\chi(\mathbf{r})}\left[\frac{dI((\mathbf{r},\tilde{\mathbf{s}})}{ds}\right]$$
$$= -I(\mathbf{r},\tilde{\mathbf{s}}) + [1-\omega(\mathbf{r})]\left(\frac{n^2\bar{\sigma}T^4}{\pi}\right)$$
$$+ \frac{\omega(\mathbf{r})}{4\pi}\oint p(\tilde{\mathbf{s}}',\tilde{\mathbf{s}})I(\mathbf{r},\tilde{\mathbf{s}}')d\Omega'. \qquad (1.2.15)$$

Equation (1.2.15) is the general form of radiative transfer equation for heat transfer in grey approximation.

It is clear from equations (1.2.8)–(1.2.12) and (1.2.15) that in the presence of thermal as well as scattering contributions to source functions, the radiative transfer equations turn out to be integro-differential equations. It may also be mentioned that the source functions can assume much more complex forms when the medium is non-homogeneous, the scattering process is non-coherent and anisotropic and the axial symmetry of the radiation field is not present.

The Radiative Transfer Equation in Different Geometries in Stationary Medium

Plane Parallel System

When the properties of the radiation field is uniform in each plane layer, say (x,y), we may represent the medium by a system of planes stratified normal to a particular axis Z (say). In this case, the medium is said to be plane parallel. To study radiative transfer in such medium, the linear distances are measured along the Z-axis, the direction normal to the planes of stratification.

22   1. Basic Notions

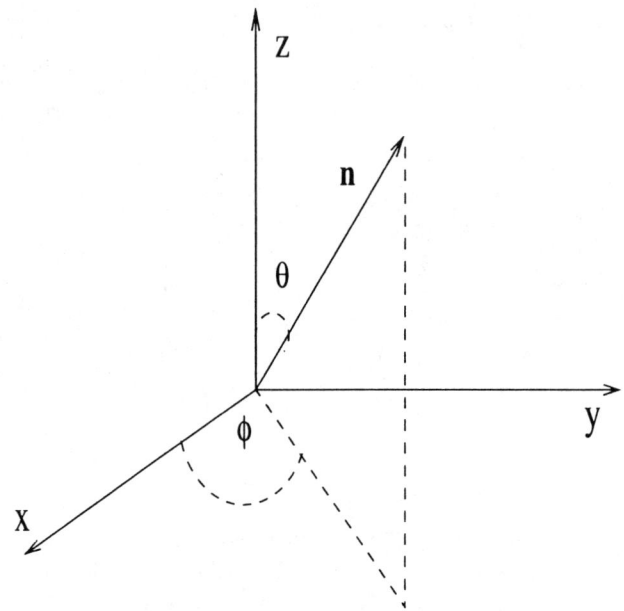

Figure 1.2. Coordinate system for plane-parallel medium

The specific intensity $I(z,t;\theta,\varphi,\nu)$ indicates that the direction of the ray $\tilde{s}$ makes an angle θ with the Z-direction and φ is the azimuth angle referred to some suitably chosen axis. Thus, (θ,φ) determines the direction of the ray at any point z.

The frequency dependence of the specific intensity can be suppressed when the model of the medium considered is grey or the specific intensity of the ray considered is monochromatic or frequency-integrated. If the radiation field is axially symmetric about the Z-axis, the specific intensity is independent of φ. Then for plane parallel or slab media

$$\frac{d}{ds} = \frac{\partial}{\partial z}\frac{dz}{ds} + \frac{\partial}{\partial t}\frac{dt}{ds} = \cos\theta\frac{\partial}{\partial z} + \frac{1}{c}\frac{\partial}{\partial t}. \qquad (1.2.16)$$

1.2 The Equation of Radiative Transfer

When the specific intensity is time independent

$$\frac{d}{ds} = \cos\theta \frac{\partial}{\partial z} = \mu \frac{\partial}{\partial z}. \quad (1.2.17)$$

Further, the optical depth τ is defined by

$$d\tau = -\chi(z,\nu)dz, \quad (1.2.18)$$

implying that

$$\tau = \int_z^\infty \chi(z,\nu)dz. \quad (1.2.19)$$

Then the transfer equation in a plane parallel, axially symmetric medium with the specific intensity $I(z,t;\mu,\nu)$ reads as

$$[\frac{1}{c}\frac{\partial}{\partial t} + \mu\frac{\partial}{\partial z} I(z,t;\mu,\nu)] = -\chi(z,t;\nu)I(z,t;\mu,\nu) + \eta(z,t;\nu) \quad (1.2.20)$$

where μ and ν are parameters, indicating the direction and frequency.

The radiative transfer equation for grey or frequency-integrated axially symmetric plane parallel medium with time independent specific intensity $I(\tau,\mu)$ is given by

$$\mu \frac{\partial I(\tau,\mu)}{\partial \tau} = I(\tau,\mu) - S(\tau). \quad (1.2.21)$$

The plane parallel atmosphere may be semi-infinite or finite depending on optical depth τ extending from $\tau = 0$ to $\tau \to \infty$ or from $\tau = 0$ to $\tau = \tau_1$. The source function $S(\tau)$ takes different forms depending on the nature of the medium interacting with the radiation.

Plane Parallel Scattering Atmospheres

For a non-absorbing, non-emitting grey atmosphere where interaction between radiation and matter takes place only by way of non-coherent scattering, the source function $S(\tau)$ can be written as

$$S(\tau) = \frac{1}{2}\int_{-1}^1 p^0(\mu,\mu')I(\tau,\mu')d\mu', \quad (1.2.22)$$

where the phase function $p^0(\mu,\mu')$ is given by

$$p^0(\mu,\varphi,\mu') = \frac{1}{2\pi}\int_0^{2\pi} p^0(\mu,\varphi;\mu',\varphi')d\varphi'. \tag{1.2.23}$$

In the axially symmetric case we may write

$$p^0(\mu,\varphi,\mu') = p^0(\mu,\mu').$$

In (1.2.22), axial symmetry of the radiation field is assumed. In the isotropic case,

$$p^0(\mu,\mu') = 1. \tag{1.2.24}$$

The equation of transfer is

$$\mu\frac{dI(\tau,\mu)}{d\tau} = I(\tau,\mu) - J(\tau). \tag{1.2.25}$$

Medium in LTE and the Milne Problem

After Eddington, it is assumed that in an atmosphere in local thermodynamic equilibrium (LTE), at each point of the medium a temperature $T(r)$ may be assigned. The properties of the medium in the immediate neighbourhood of the point r is similar to that confined to an ideal adiabatic enclosure at temperature $T(r)$. In this case, Kirchoff's law holds good and

$$\frac{\eta((r,\nu)}{\chi(r,\nu)} = B_\nu(T), \tag{1.2.26}$$

where $B_\nu(T)$ is the Planck function [cf. (1.1.60)].

In plane parallel medium in LTE, the source function $S_\nu(\tau)$ is given by

$$S_\nu(\tau) = B_\nu(T), \tag{1.2.7}$$

and the radiative transfer equation in the case of non-grey plane parallel medium in LTE reads as

$$\mu\frac{dI(\tau,\mu,\nu)}{d\tau} = I(\tau,\mu,\nu) - B_\nu(T) \tag{1.2.28}$$

and the corresponding grey equation is

$$\mu\frac{dI(\tau,\mu)}{d\tau} = I(\tau,\mu) - B(T). \tag{1.2.29}$$

Liasion Condition

Further if it is assumed that the frequency-integrated energy of radiation is conserved from layer to layer, the medium is said to be in "strict radiative equilibrium". This liasion condition for non-grey plane parallel medium can be written as

$$\int_0^\infty \chi(\tau,\nu)J(\tau,\nu)d\nu = \int_0^\infty \chi(\tau,\nu)B_\nu(T)d\nu, \qquad (1.2.30.)$$

where, the mean intensity,

$$J(\tau,\nu) = \frac{1}{2}\int_{-1}^1 I(\tau,\mu,\nu)d\mu.$$

The transfer equation (1.2.28) and the liasion condition (1.2.30) taken together imply conservation of flux, that is,

$$F = 2\int_{-1}^1 I(r,\mu,\nu)\mu d\mu = \text{constant.} \qquad (1.2.31)$$

For grey approximation, taking

$$I(\tau,\mu) = \int_0^\infty I(\tau,\mu,\nu)d\nu \quad \text{and} \quad B(T) = \int_0^\infty B_\nu(T)d\nu, \quad (1.2.32)$$

the liasion condition becomes $J(\tau) = B(\tau)$ and the radiative transfer equation is still given by (1.2.29).

Milne Problem

The problem described in (1.2.25)–(1.2.32) (or its purely scattering analogue) was first introduced by Milne for analysing solar spectrum. The problem is known as Milne problem when the boundary conditions under which it is solved are:

(a) absence of external radiation at the free surface of the medium

$$I(0,-\mu) = 0. \quad \text{for} \quad 0 < \mu \le 1; \qquad (1.2.33)$$

and

(b) absence of singularity as $\tau \to \infty$ (assuming the medium to be semi-infinite), i.e.

$$B(\tau)\exp(-\tau) \to 0 \quad \text{as} \quad \tau \to \infty. \tag{1.2.34}$$

The formal solution of Milne type transfer problem in grey, semi-infinite, plane parallel slab under the boundary conditions (1.2.33) and (1.2.34) is given by

$$I_+(\tau,\mu) = \int_\tau^\infty B(t)\exp\left(\frac{-(t-\tau)}{\mu}\right)\frac{dt}{\mu}; \quad 0 < \mu \leq 1$$

$$I_-(\tau,-\mu) = \int_0^\tau B(t)\exp\left(\frac{-(\tau-t)}{\mu}\right)\frac{dt}{\mu}; \quad 0 < \mu \leq 1. \tag{1.2.35}$$

In the study of purely scattering medium, $S(\tau) = J(\tau)$ and its formal solution can be obtained in an identical way. In passing it may be mentioned that, for a finite atmosphere, the incident radiation at both the bounding surfaces are to be specified.

The radiative transfer equations introduced above are solved under appropriate boundary conditions for the knowledge of specific intensity and source function. The methodologies for solutions of transfer problems in plane parallel participating media are in a very advanced state of development. Excellent accounts of them are given in the books of Chandrasekhar (1950), Kourganoff (1952), Sobolev (1962), Busbridge (1970), Mihalas (1978), Özisik (1973), Davison and Sykes (1958), Case and Zweifel (1967) and others.

1.3 Radiative Transfer in Stationary Spherical Medium

In this book, we are interested in studying the radiative transfer problems for a class of stars whose abnormal nature of spectrum suggests that their spectral structure is characteristic of an extended moving stellar atmosphere. The radiative transfer equation appropriate to these problems should be written in spherical geometry and that too in moving media. However, for completeness and historical interest, we have outlined the derivation of radiative transfer equation in plane parallel medium in (1.2). In this section, we shall consider the derivation of radiative transfer equation in spherically symmetric stationary media. The transfer equation in spherically symmetric moving media will be established later in chapter 2. The subsequent chapters will be devoted to the study of such transfer equations.

Equations of Radiative Transfer in a Spherically Symmetric Stationary Atmosphere

The integro-differential form of the equation of radiative transfer through a spherical envelope surrounding a core star will be deduced.

In the general form of radiative transfer equation given in (1.2.3)–(1.2.5), the space-derivative (d/ds) is expressed in spherical polar coordinates (r, θ, φ). The radiation field is assumed to be azimuth independent. The intensity $I(\mathbf{r}, t; \tilde{\mathbf{s}}, \nu)$ is written as

$$I(\mathbf{r}, t; \tilde{\mathbf{s}}, \nu) = I(\mathbf{r}, t; \theta, \nu), \qquad (1.3.1)$$

where ν is the frequency parameter. Then

$$\frac{dI(r, t; \theta, \nu)}{ds} = \frac{\partial I}{\partial r}\frac{dr}{ds} + \frac{\partial I}{\partial \theta}\frac{d\theta}{ds} + \frac{\partial I}{\partial t}\frac{dt}{ds}. \qquad (1.3.2)$$

From Fig. (1.3), we have

$$dr = ds \cos\theta, \quad rd\theta = -ds \sin\theta, \quad cdt = ds. \qquad (1.3.3)$$

Then

$$\frac{dI(r, t; \theta, \nu)}{ds} = \cos\theta \frac{\partial I(r, t; \theta, \nu)}{\partial r} - \frac{\sin\theta}{r}\frac{\partial I(r, t; \theta, \nu)}{\partial r}$$
$$+ \frac{1}{c}\frac{\partial I(r, t; \theta, \nu)}{\partial t}. \qquad (1.3.4)$$

Now writing $\cos\theta = \mu$, we have $I(r, t; \theta, \nu) = I(r, t; \mu, \nu)$. Hence

$$\frac{dI(r, t; \mu, \nu)}{ds} = \frac{\partial I}{\partial r}\frac{dr}{ds} + \frac{\partial I}{\partial \mu}\frac{d\mu}{ds} + \frac{\partial I}{\partial t}\frac{dt}{ds}. \qquad (1.3.5)$$

Since $\cos\theta = \mu$,

$$\frac{d\mu}{ds} = \frac{d\mu}{d\theta}\cdot\frac{d\theta}{ds} = -\sin\theta\left(\frac{-\sin\theta}{r}\right) = \frac{\sin^2\theta}{r} = \frac{1-\mu^2}{r}. \qquad (1.3.6)$$

Then the equation of radiative transfer in spherical geometry for diffuse radiation in spherically symmetric atmospheres can be written as [cf. equations (1.2.3)–(1.2.5); (1.3.2)–(1.3.6)]

$$\frac{1}{c}\frac{\partial I(r,t;\mu,\nu)}{\partial t} + \mu\frac{\partial I(r,t;\mu,\nu)}{\partial r} + (\frac{1-\mu^2}{r})\frac{\partial I(r,t;\mu,\nu)}{\partial \mu}$$
$$= -\chi(r,t;\nu)[(I(r,t;\mu,\nu) - S(r,t;\mu,\nu)], \qquad (1.3.7)$$

where $S(r,t;\mu,\nu)$ is the source function. For time-independent specific intensity $I(r,\mu,\nu)$, the radiative transfer equation reads as

$$\mu\frac{\partial I(r,\mu,\nu)}{\partial r} + (\frac{1-\mu^2}{r})\frac{\partial I(r,\mu,\nu)}{\partial \mu}$$
$$= -\chi(r,\nu)[(I(r,\mu,\nu) - S(r,\mu,\nu)], \qquad (1.3.8)$$

with

$$S(r,\mu,\nu) = \frac{\omega(r)}{2}\int_0^\infty \frac{R(\nu',\nu)}{\phi(\nu)}d\nu'\int_{-1}^1 I(r,\mu',\nu)d\mu' + g(r,\mu). \qquad (1.3.9)$$

In equations (1.3.8) and (1.3.9), we may write

$$\chi(r,\nu) = \bar{\chi}(r)\phi(\nu), \qquad (1.3.10)$$

where $\phi(\nu)$ is the absorption profile. $R(\nu',\nu)$ is the angle averaged redistribution function. For the frequency independent intensity

$$\int_0^\infty \frac{R(\nu',\nu)}{\phi(\nu)}d\nu' = \frac{\psi(\nu)}{\phi(\nu)}$$

gives the ratio of the emission profile $\psi(\nu)$ to the absorption profile $\phi(\nu)$.

The first term on the right hand side of (1.3.9) is the contribution from scattering process to the source function and the second term is the contribution to the source function from sources other than scattering.

1.3 Radiative Transfer in Stationary Spherical Medium

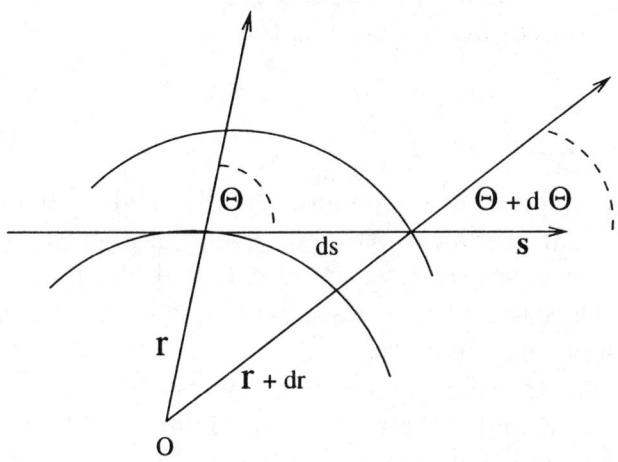

Figure 1.3. The r-θ and s coordinates for spherical medium

The source function for a coherent scattering frequency independent radiation field is given by

$$S(r,\mu) = \frac{\omega(r)}{2} \int_{-1}^{1} I(r,\mu')p(\mu,\mu')d\mu' + B_0(r) + B_1(r), \quad (1.3.11)$$

and the corresponding radiative transfer equation is

$$\mu \frac{\partial I(r,\mu)}{\partial r} + \frac{1-\mu^2}{r}\frac{\partial I(r,\mu)}{\partial \mu} = -\chi(r)[(I(r,\mu) - S(r,\mu)]. \quad (1.3.12)$$

In (1.3.11) and (1.3.12), $\omega(r)$ is the albedo for single scattering where $0 \leq \omega \leq 1$, and $p(\mu,\mu')$, is the phase function for scattering. $B_0(r)$ is the contribution to the source function from the reduced flux incident on the bounding surfaces and $B_1(r)$, that from internal sources other than scattering.

Boundary Conditions

In spherically symmetric media, we usually deal with problems where a point source or an emitting core and/or an absorbing core is surrounded by a spherical or spherical shell envelope. In most of these problems the proper statement of boundary conditions does not present any difficulty.

The radiative transfer equations of types (1.3.7), (1.3.8) and (1.3.12) are solved under the relevant boundary conditions. Some such boundary conditions are stated below.

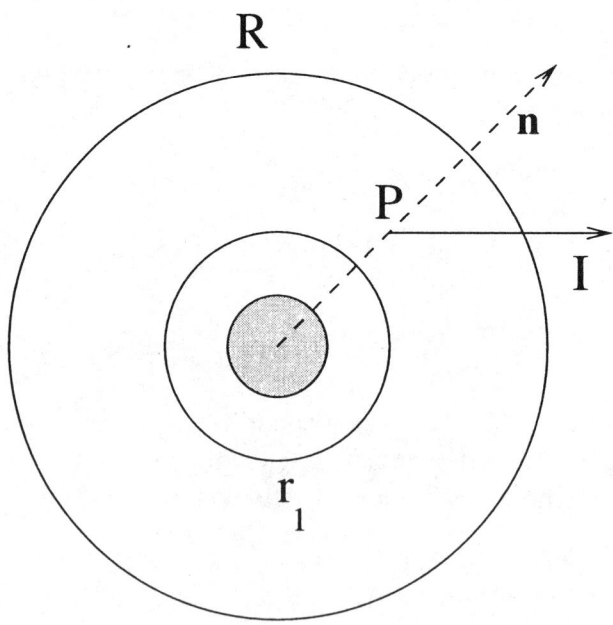

Figure 1.4. Spherical shell surrounding a core

A Sphere of Finite Radius Surrounding a Point Source of Energy

We write the outgoing and incoming radiation at any point as $I(r, +\mu, \nu)$ and $I(r, -\mu, \nu)$ respectively. The boundary condition at the outer surface $r = R$ of the sphere surrounding the point source can be written as absence of any external incident radiation on the free surface. That is

$$I(R, -\mu, \nu) = 0. \tag{1.3.13}$$

As the lower boundary condition, we must avoid any singularity at the source point $r = 0$.

A Spherical Shell Medium Surrounding a Core

We suppose a spherical shell bounded by surfaces of radii r_1 and R ($r_1 < r < R$) surrounds a partially or perfectly absorbing or reflecting core. The boundary conditions for solving the transfer problems for these cases can be written as

$$\begin{aligned} I(R, -\mu, \nu) &= 0, \\ I(r_1, +\mu, \nu) &= \gamma\, I(r_1, -\mu, \nu) \end{aligned} \tag{1.3.14}$$

for
$$0 \leq \nu \leq \infty,\ 0 < \mu \leq 1,\ 0 \leq \gamma \leq 1.$$

Here
$$\gamma = 0 \ \Rightarrow\ \text{a perfectly absorbing core},$$

and
$$\gamma = 1 \ \Rightarrow\ \text{a perfectly reflecting core}.$$

It should be noted that the boundary conditions (1.3.13) and (1.3.14) refer only to diffuse radiation. If the source function contains sources other than scattering, their contributions have to be taken into account in spelling out the boundary conditions for the total intensity.

1.4 Non-LTE Conditions and Statistical Equilibrium in Line Transfer Problems

In most of the earlier sections, we have assumed that the physical conditions in participating media can be approximated by local thermodynamic equilibrium (LTE). The pre-requisite for the existence of this condition is that the particle distribution function is Maxwellian and thermodynamic equilibrium prevails. Every transition between atomic levels is cancelled by an exactly inverse transition or that the principle of detailed balance holds good in all processes. When the particle distribution is Maxwellian, the collision processes occur at the local kinetic temperatures at their equilibrium rates. Thus when the excitation and de-excitation of atomic levels take place by inelastic collisions alone, we consider the situation to fall within the perview of LTE.

In general, both radiative and collision processes contribute to the excitation and de-excitation of atomic levels. When the radiation field is Planckian, the principle of detailed balancing still holds good and leads the medium to LTE conditions. In the far interior of the stars, the conditions are compatible for LTE approximations. However, near the surface the radiative rates are not detailed balanced and the system is driven away from LTE conditions. Here non-LTE situation prevails.

However, in presence of both radiative and collisional processes taking part in the excitation and de-excitation of atomic levels, the rate coefficients can still be found by detailed balancing. The excitation and de-excitation processes are to be jointly detailed balanced. The relevance of LTE assumption depends on the relative contributions of collisional excitation and ionisation rates and radiative excitation and ionisation rates. When the former is predominant LTE prevails while in the opposite case we are led to non-LTE. In many cases of the study of radiative transfer in stellar atmospheres, this departure from LTE conditions cannot be ignored without incurring serious errors.

Statistical Equilibrium Equations

The equation of statistical equilibrium is an expression that states the equilibrium between the various processes leading to the establishment of an equilibrium state.

We shall assume that the state of the gas is specified by its kinetic temperature, the degrees of excitation and ionisation of each atomic level. The gas is also supposed to be in a steady state. We consider the main types of elementary processes populating into or de-populating from an energy level i of atoms of particular species.

We take a volume element in a moving media and consider the number of particles of a given state i and of particular kind x will change in time. The net rate at which particles from state j are brought to state i by radiative and collisional processes is given by

$$(\frac{\partial n_{ix}}{\partial t}) = \sum_{j \neq i}(n_{jx}P^x_{ji} - n_{ix}P^x_{ij}) - \nabla \cdot (n_{ix}\mathbf{V}), \qquad (1.4.1)$$

where $\mathbf{V}$ is the velocity of the moving media.

P_{ji} denotes the total rate from level j to i. The second term on the right hand side gives the total number of particles entering and leaving the volume element. The total number of particles of type x

$$N_x = \sum_i n_{ix}.$$

Hence the continuity equation reads as

$$(\frac{\partial N_x}{\partial t}) + \nabla \cdot (N_x \mathbf{V}) = 0. \qquad (1.4.2)$$

If m_x is the mass of each particle of type x, multiplying (1.4.2) by m_x and summing over all species of particles in the volume element, we have

$$\frac{\partial \rho}{\partial t} + \nabla \cdot (\rho \mathbf{V}) = 0, \quad \rho = \sum_x m_x N_x. \qquad (1.4.3)$$

For a steady flow

$$\sum_{j \neq i}(n_{jx}P^x_{ji} - n_{ix}P^x_{ij}) = \nabla \cdot (n_{ix}\mathbf{V})$$

For steady state in static atmosphere we have

$$n_i \sum_{j \neq i} P_{ij} - \sum_{j \neq i} n_j P_{ji} = 0. \qquad (1.4.4)$$

In what follows we shall consider only the static medium and the radiative and collisional transitions between bound states.

We first list the modes and number of transitions to level i per cm^3 per sec.

(a) Spontaneous and stimulated radiative transitions from higher discrete levels:

$$\sum_{k=i+1}^{\infty} n_k (A_{ki} + B_{ki} \bar{J}_{ik})$$

where n_k is the population of level k.

A_{ki} is the Einstein coefficient for spontaneous emission.
B_{ki} is the Einstein coefficient for stimulated emission.
$\bar{J}_{ik}$ is the mean intensity of the radiation in the $i \to k$ line weighted with the profile of the absorption coefficient [cf. (3.1 11)].

(b) Spontaneous and radiative recombination:

$$n_e n^+ (A_{ci} + B_{ci} \bar{J}_{ic})$$

where

n_e is the electron density.
n^+ is the ionic concentration.
A_{ci} is the number of recombinations onto the level i per sec per ion and unit electron density.
$B_{ci} \bar{J}_{ic}$ is the number of stimulated photon recombinations to level i per sec per ion and unit electron density.

(c) Collision of second kind (collision induced transition from upper levels):

$$n_e \sum_{k=i+1}^{\infty} n_k C_{ki}$$

1.4 Non-LTE Conditions and Statistical Equilibrium

where

C_{ki} is the number of $k \to i$ transitions induced by electron collisions per sec per atom in level k and per unit electron density.

(d) Three body recombination:
$$n_e^2 n^+ C_{ci}$$

where

C_{ci} is the number of three body recombinations into level i per sec per ion and unit electron density.

(e) Photo-excitation from lower levels:
$$\sum_{j=1}^{i-1} n_j B_{ji} \bar{J}_{ji}.$$

(f) Collisional excitation:
$$n_e \sum_{j=1}^{i-1} n_j C_{ji}.$$

Therefore the net transition into i per cm^3 per sec is the sum of (a) to (f).

$$\sum_{k=i+1}^{\bar{C}} n_k (A_{ki} + B_{ki} \bar{J}_{ik} + n_e C_{ki})$$
$$+ \sum_{j=1}^{i-1} n_j (B_{ji} \bar{J}_{ji} + n_e C_{ji}) \quad (1.4.5)$$

where $\bar{C}$ denotes the continuum state. In (1.4.5) we have excluded the contributions of (b) and (e).

Now we take note of the processes effecting transitions from level i.

(a') Spontaneous and stimulated radiative transitions to lower levels:
$$n_i \sum_{j=1}^{i-1} (A_{ij} + B_{ij} \bar{J}_{ji}).$$

(b') Transition downwards induced by collision of second kind (de-activation):
$$n_e n_i \sum_{j=1}^{i-1} C_{ij}.$$

(c') Photo-excitation into higher levels:
$$n_i \sum_{k=i+1}^{\infty} B_{ik} \bar{J}_{ik}.$$

(d') Upward transition due to collision with electrons:
$$n_e n_i \sum_{k=i+1}^{\infty} C_{ik}.$$

(e') Photo-ionisation:
$$n_i B_{ic} \bar{J}_{jc}.$$

(f') Ionisation by electron impact:
$$n_e n_i C_{ic}.$$

The total number of transitions from i is the sum of (a') to (f') (excluding the contributions of (e') and (f'))

$$n_i \sum_{j=1}^{i-1} (A_{ij} + B_{ij}\bar{J}_{ji} + n_e C_{ij})$$
$$+ n_i \sum_{k=i+1}^{\bar{C}} (B_{ik}\bar{J}_{ik} + n_e C_{ik}). \tag{1.4.6}$$

Now equating (1.4.5) and (1.4.6), we get the following system of equations for the population n_i as

$$\sum_{k=i+1}^{\bar{C}} n_k(A_{ki} + B_{ki}\bar{J}_{ik} + n_e C_{ki}) + \sum_{j=1}^{i-1} n_j(B_{ji}\bar{J}_{ji} + n_e C_{ji})$$
$$= n_i \sum_{j=1}^{i-1} (A_{ij} + B_{ij}\bar{J}_{ji} + n_e C_{ij}) + n_i \sum_{k=i+1}^{\bar{C}} (B_{ik}\bar{J}_{ik} + n_e C_{ik}). \tag{1.4.7}$$

We can determine the state of the gas at any point of the medium from (1.4.7), when the mean intensity $\bar{J}$ of the radiation field,

1.4 Non-LTE Conditions and Statistical Equilibrium

temperature, and electron density n_e are given at that point. It may be mentioned here that (1.4.7) has been set up for a very simple model. For the detailed account of the construction of rate equations the readers are referred to Mihalas and Mihalas [1984 p. 386–398].

The system of algebraic equations (1.4.7) are the statistical equilibrium equations, the solution of which is a simple routine. We consider a two-level atom which contains bound levels and a continuum and seek to obtain an analytical expression for source function. We shall pay attention to the appearance of scattering term in the expression for source function.

The Line Source Function for a Two Level Atom

We designate the upper level by u and lower level by l. To derive the expression for source function, we assume the line to be strong and consider the absorption coefficient and emissivity of the line itself. We invoke the relations between the Einstein coefficients A_{ul}, B_{lu}, B_{ul} through the principle of detailed balance and obtain the expressions for absorption coefficient (opacity) χ and emissivity η [cf. Mihalas and Mihalas (1984) p. 330].

$$g_u B_{ul} = g_l B_{lu} \tag{1.4.8}$$

$$A_{ul} = (2h\nu_{lu}^3/c^2) B_{ul}, \tag{1.4.9}$$

where g_u and g_l are statistical weights and $h\nu_{lu}$ is equal to the energy difference between the two levels measured relative to the ground state.

The line absorption coefficient

$$\chi_l(\nu) = n_l(B_{lu} h\nu_{lu}/4\pi)[1 - (\frac{g_l n_u}{g_u n_l})] \phi_{\nu_0}, \tag{1.4.10}$$

$$\eta_l(\nu) = n_u (A_{ul} h\nu_{lu}/4\pi) \phi_{\nu_0}. \tag{1.4.11}$$

In (1.4.10) and (1.4.11), the line profile is expressed in comoving frame. When the opacity and emissivity are defined in lab frame, Doppler shifts in ϕ_{ν_0} have to be taken into account and ϕ_{ν_0} will mean

$$\phi_{\nu_0}(\mathbf{r}, t; \mathbf{n}_0, \nu_0) = \phi[\mathbf{r}, t; \mathbf{n}, \nu(1 - \mathbf{n} \cdot \mathbf{V}/c)]. \tag{1.4.12}$$

The source function S_l is given by

$$S_l = \frac{n_u A_{ul}}{n_l B_{lu} - n_u B_{ul}}$$
$$= \frac{2h\nu^3/c^2}{[(\frac{g_u n_l}{g_l n_u}) - 1]}. \qquad (1.4.13)$$

From (1.4.7), we may write for the populations n_u and n_l (when the bound free transitions are excluded)

$$n_l[B_{lu}\bar{J} + C_{lu}] = n_u[A_{ul} + B_{ul}\bar{J} + C_{ul}]. \qquad (1.4.14)$$

The collision rates C_{ul} and C_{lu} per atom in the initial state are related by

$$\frac{C_{lu}}{C_{ul}} = (\frac{g_u}{g_l}) \exp(-h\nu_{ul}/kT). \qquad (1.4.15)$$

Then from (1.4.8), (1.4.9), (1.4.13), (1.4.14) and (1.4.15) we may write

$$S_l = \frac{\bar{J} + \epsilon' B^s}{1 + \epsilon'}, \qquad (1.4.16)$$

where

$$\epsilon' = (\frac{C_{ul}}{A_{ul}})[1 - \exp(-h\nu/kT)]. \qquad (1.4.17)$$

In this case B^s is the Planck function defined in (1.1.60)

For multilevel atoms when bound-free transitions are included ϵ' and B^s will involve additional quantities, but the form of the equation (1.4.16) is unaltered.

In stellar atmospheres, transitions are in general induced by their own radiation field. In order to solve radiative transfer problems in stellar envelopes, concurrent solutions of radiative transfer and statistical equilibrium equations are necessary particularly under non-LTE conditions. This implies that the physical condition at any particular point of the atmosphere is built up from the physical state of the entire medium.

Thermodynamic Equilibrium

The physical state of a system in thermodynamic equilibrium is completely determined by the temperature T. The velocity distribution of the particles constituting the system is Maxwellian. The degree of ionisation and excitation of the particles is given by Saha's equation and Boltzmann equation. The principle of detailed balancing holds good for every process, implying that the number of radiative transition $i \to j$ is completely compensated by photo-excitation $j \to i$. (i is the upper level and j, the lower in the two level transition.)

Then

$$n_i[A_{ij} + B_{ij}(\frac{c}{4\pi})u_{\nu_{ji}}] = n_j B_{ji}(\frac{c}{4\pi})u_{\nu_{ji}}; \quad j < i, \ i = 2, \ldots$$
(1.4.18),

where $u_{\nu_{ji}}$ is the Planckian radiation density of frequency ν_{ji} given by

$$\begin{aligned} u_{\nu_{ji}}(T) &= \frac{8\pi h \nu_{ji}^3}{c^3}[\exp(h\nu_{ji}/kT) - 1]^{-1} \\ &= \frac{4\pi}{c} B_{\nu_{ji}}(T), \end{aligned}$$
(1.4.19)

where $B_{\nu_{ji}}(T)$ is the Planck function.

Similarly, the radiative ionisation from i is balanced by radiative recombination into i. This implies

$$n_e[A_{ci} + B_{ci}(\frac{c}{4\pi})u_{\nu_{ic}}] = n_i B_{ic}(\frac{c}{4\pi})u_{\nu_{ic}}; \quad i = 1, 2, \ldots \quad (1.4.20)$$

The detailed balanced relations for collisional transitions can be written as

$$n_i C_{ij} = n_j C_{ji} \quad i, j = 1, 2, \ldots, i \neq j, \quad (1.4.21)$$

$$n_e n^+ C_{ci} = n_i C_{ic} \quad i = 1, 2, \ldots . \quad (1.4.22)$$

It may be emphasised here that the existence of LTE conditions depends on the relative importance of collisional transitions over radiative transitions. In fact, a sufficient condition for a strict LTE assumption to be valid is that the radiative transition is negligible compared to the collisional transition. We pay attention to this in using non-LTE assumption in transfer problems for stellar atmospheres.

1.5 Outline of Methods of Solution of Radiative Transfer Equation in Spherically Symmetric Static Medium

The general form of transfer equation is given in equation (1.2.3) as

$$\frac{1}{\chi(\mathbf{r},\nu)} \frac{dI(\mathbf{r},\tilde{\mathbf{s}},\nu)}{ds} = -I(\mathbf{r},\tilde{\mathbf{s}},\nu) + S(\mathbf{r},\nu), \qquad (1.5.1)$$

where $\chi(\mathbf{r},\nu)$, $\eta(\mathbf{r},\nu)$ have been defined by (1.2.4) and (1.2.5) and

$$S(\mathbf{r},\nu) = \frac{\eta(\mathbf{r},\nu)}{\chi(\mathbf{r},\nu)}.$$

The formal solution of (1.5.1) can be written out integrating it along the path of the ray in the direction $\tilde{\mathbf{s}}$ under the boundary condition that the specific intensity of the radiation at the entry point $s = s_0$ is given. That is for the boundary condition at s_0, we write

$$I(s,\tilde{\mathbf{s}},\nu) = I_{0\nu} \quad \text{at} \quad s = s_0. \qquad (1.5.2)$$

The formal integration of the transfer equation (1.5.1) under (1.5.2) leads us to

$$\begin{aligned}I(s,\tilde{\mathbf{s}},\nu) &= I_{0\nu} \exp\left[-\int_{s_0}^{s} \chi(s',\nu)ds'\right] \\ &+ \int_{s_0}^{s} [\,\chi(s',\nu)S(s',\tilde{\mathbf{s}},\nu)\exp\left[-\int_{s_0}^{s'} \chi(s'',\nu)ds''\right]]ds'. \end{aligned} \qquad (1.5.3)$$

Equation (1.5.3) is the integral form of transport equation. The first term on the right hand side accounts for the reduced incident radiation from $s = s_0$ to s and the second term gives the measure of the diffuse radiation. However, (1.5.3) is not the solution of (1.5.1) since the source function $S(s',\tilde{\mathbf{s}},\nu)$ may contain the unknown specific intensity.

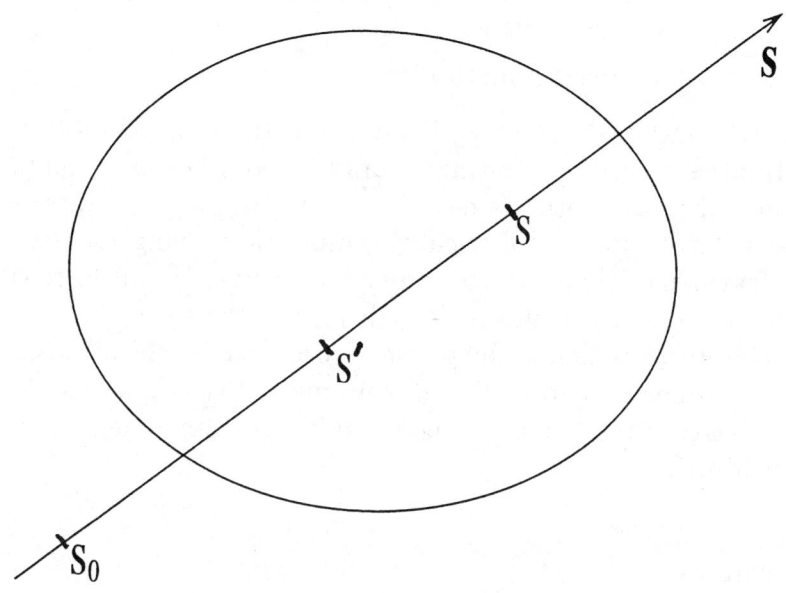

Figure 1.5. Direction of ray through an arbitrary medium

1.6 Basic Methods of Solution of Transfer Problems in Spherically Symmetric Static Media

The basic methods for solving integro-differential and integral equations of radiative transfer in spherically symmetric stationary media are in a very advanced state of development. Broadly speaking they can be classified into four main categories:

(i) moment or moment-like methods,

(ii) conversion of two point boundary value problems into initial value problems,

(iii) methods involving solution of integral equations of transfer,

(iv) strictly numerical methods.

A self contained account of the use of these methods is given in a treatise by the present authours [Sen and Wilson (1990)].

Some of these methods have been extended to the solution of radiative problems in spherically symmetric moving media. The basic features of them will be explained in Part II and Part III of this book along with simple illustrations of their use.

In chapter 2 we define the physical properties of the atmosphere and the radiation field in the moving media. We shall also derive the radiative transfer equations in lab frame, comoving frame and mixed frame.

References

Busbridge, I.W. (1960): The Mathematics of Radiative Transfer. Cambridge University Press, Cambridge.
Case, K.M., Zweifel, P.F. (1967): Linear Transport Theory. Addison Wesley, Massachusetts.
Chandrasekhar, S. (1960): Radiative Transfer. Clarendon Press, Oxford.
Davison, B., Sykes, J.B. (1958): Neutron Transport Theory. Clarendon Press, Oxford.
Kourganoff, V. (1952): Basic Methods in Transfer Problems. Clarendon Press, Oxford.
Mihalas, D. (1978): Stellar Atmospheres. W. Freeman and Co., San Francisco.
Mihalas, D., Mihalas, B.W. (1984): Foundations of Radiation Hydrodynamics. Oxford University Press, New York.
Ozisik, M. N. (1973): Radiative Transfer. Wiley Interscience Publications, New York.
Sen, K.K., Wilson, S J. (1990): Radiative Transfer in Curved Media. World Scientific, Singapore.
Sobolev, V. V. (1962): A Treatise on Radiative Transfer. Van Nostrand, New York.

Chapter 2. Radiation Field and Transfer Equations in Moving Media

The atmospheres of stars consist of compressible dilute gas made up of atoms, ions and free electrons. In the case of atmospheres in motion, radiative force is one of the factors which causes motion. The radiative transfer through spherically symmetric moving media is our subject of study. We shall examine here how the physical quantities describing the radiative field and material are modified due to expansion, rotation or a combination of both. The moments of specific intensities will be calculated for the moving media and the radiative transfer equations in lab, mixed and comoving frames will be deduced. Be-, P-Cygni, W-Rayet stars, supernovae and such celestial objects are all endowed with extensive atmospheres. The abnormal spectra of these stars are explained in terms of macroscopic motion in their envelope. The radiative transfer problems in these atmospheres can be approximated by considering them to be that in spherically symmetric moving media.

When the radiative transfer equation for stationary spherically symmetric envelope is written down, the opacity and emission coefficients in the lab frame are isotropic or angle independent. However, in the case of moving stellar envelope, Doppler shift, aberration and advection make both opacity and emissivity angle dependent and hence anisotropic. To solve radiative transfer equation in a moving medium in the lab frame (a frame at rest with respect to the centre of the star), attention has to be paid to this aspect of the problem.However, this difficulty can be avoided by expressing the transfer equation in a frame planted on the moving fluid element under observation and comoving with it. We shall confine ourselves to the study of radiative transfer through the spherically symmetric envelopes of the stars,which exhibit the presence of velocity gradients along the line of sight of the observer. No attempt will be made to study the micro- or macro-turbulence velocity fields.

To formulate the appropriate equations for the study of transport phenomenon in a moving medium, account should be taken of the dynamical equations of fluid flow in addition to the equations of radiative transfer. However, we shall study only the radiative transfer equations in moving media and not deal with the fluid flow equations. Special attention has to be paid to the choice of frame of reference in expressing the transfer equations and measuring the radiative and material quantities. Judicious choice of the frames may give considerable computational advantage.

We shall consider the transfer equations in three distinct frames.

(a) Inertial or Lab frame:

Here the observer is fixed with respect to the centre of the core star around which the atmosphere is in motion.

(b) Mixed frame:

Here $V/c \ll 1$, and a first order expansion of χ and η about inertial frame frequency ν is made. This takes care of anisotropy of χ and η.

(c) Comoving frame:

Here every element of fluid contributes to the set of inertial frames, each of which has a velocity instantaneously coinciding with that of the fluid element.

2.1 Radiative Transfer Equations in Inertial or Lab Frame for Spherically Symmetric Moving Media

Let the material of the atmosphere surrounding a core star move with a velocity $V(r)$ relative to an observer at rest in the lab frame. Between the lab frame and the system of frames planted with elements of the medium at every point r, there will be shift of frequency due to Doppler effect, aberration and advection. All three effects are of order $V(r)/c$. The frequency change in the line profile $\Delta\nu$ becomes important when $(\Delta\nu/\Delta\nu_D) = (V(r)/v_{th})$ is significant and not when $(\Delta\nu/\Delta\nu_D) = (V(r)/c)$. Here $\Delta\nu_D$ is the Doppler width $\nu_0 v_{th}/c$, where ν_0 is line centre frequency and v_{th} is the thermal velocity parameter. We assume now that the contributions of aberration and advection are small compared to that of Doppler effect. Then if ν is the frequency in the lab frame and ν', the frequency at which a photon moving in the direction $\tilde{s}$ is emitted or absorbed, (i.e. in the moving frame) we have

$$\nu' = \nu - \nu_0(\tilde{s} \cdot \mathbf{V}/c). \tag{2.1.1}$$

It is convenient sometimes to state the frequency in dimensionless form as

$$x = \frac{\nu - \nu_0}{\Delta\nu_D} \quad \text{where} \quad \Delta\nu_D = \frac{\nu_0 v_{th}}{c}.$$

For simplicity we shall focus attention on one dimensional flows only and write the radiative transfer equation in the lab frame as

$$(\frac{1}{c}\frac{\partial}{\partial t} + \tilde{s} \cdot \nabla)I(\mathbf{r},t;\tilde{s},\nu) = \eta(\mathbf{r},t;\tilde{s},\nu) - \chi(\mathbf{r},t;\tilde{s},\nu)I(\mathbf{r},t;\tilde{s},\nu). \tag{2.1.2}$$

In spherically symmetric moving media in polar coordinates (r,θ,φ), it reads as

$$\left[\frac{1}{c}\frac{\partial}{\partial t} + \mu\frac{\partial}{\partial r} + \frac{1-\mu^2}{r}\frac{\partial}{\partial \mu}\right]I(r,t;\mu,\nu)$$
$$= \eta(r,t;\mu,\nu) - \chi(r,t;\mu,\nu)I(r,t;\mu,\nu), \tag{2.1.3}$$

where $\cos^{-1}\mu$ is the angle between the photon direction and the radius vector at the point of observation. In (2.1.2) and (2.1.3) note should be taken of the fact that opacity χ and emissivity η are anisotropic and angle-frequency dependent.

Moment Equation

We obtain the zeroth and first angular moment equations of (2.1.2) by successively multiplying (2.1.2) by $(d\Omega/4\pi)$ and $\tilde{s}(d\Omega/4\pi)$

respectively and integrating over complete solid angle. Then we have for the zeroth order monochromatic moment equation

$$\frac{4\pi}{c}\frac{\partial}{\partial t}J_\nu(\mathbf{r},t;\nu) + \nabla\cdot\mathbf{F}(\mathbf{r},t;\nu)$$
$$= \oint[\eta(\mathbf{r},t;\tilde{\mathbf{s}},\nu) - \chi(\mathbf{r},t;\tilde{\mathbf{s}},\nu)I(\mathbf{r},t;\tilde{\mathbf{s}},\nu)]d\Omega. \quad (2.1.4)$$

Remembering that the monochromatic energy density E_ν is given by $E_\nu = (4\pi/c)J_\nu$, equation (2.1.4) becomes

$$\frac{\partial E_\nu}{\partial t} + \nabla\cdot\mathbf{F}_\nu = \oint[\eta(\mathbf{r},t;\tilde{\mathbf{s}},\nu) - \chi(\mathbf{r},t;\tilde{\mathbf{s}},\nu)I(\mathbf{r},t;\tilde{\mathbf{s}},\nu)]d\Omega. \quad (2.1.5)$$

Similarly the first angular moment equation can be written as

$$\frac{1}{c^2}\frac{\partial F_\nu}{\partial t} + \nabla\cdot\mathcal{P}_\nu$$
$$= \frac{1}{c}\oint[\eta(\mathbf{r},t;\tilde{\mathbf{s}},\nu) - \chi(\mathbf{r},t;\tilde{\mathbf{s}},\nu)I(\mathbf{r},t;\tilde{\mathbf{s}},\nu)]\tilde{\mathbf{s}}d\Omega. \quad (2.1.6)$$

Finally integrating (2.1.5) and (2.1.6) over all frequencies, we have the following two equations:

$$\frac{\partial E(\mathbf{r},t,\nu)}{\partial t} + \nabla\cdot\mathbf{F}(\mathbf{r},t)$$
$$= \int_0^\infty d\nu \oint d\Omega[\eta(\mathbf{r},t;\tilde{\mathbf{s}},\nu) - \chi(\mathbf{r},t;\tilde{\mathbf{s}},\nu)I(\mathbf{r},t;\tilde{\mathbf{s}},\nu)] \quad (2.1.7)$$

and

$$\frac{1}{c^2}\frac{\partial F(\mathbf{r},t)}{\partial t} + \nabla\cdot\mathcal{P}(\mathbf{r},t)$$
$$= \int_0^\infty d\nu \oint d\Omega[\eta(\mathbf{r},t;\tilde{\mathbf{s}},\nu) - \chi(\mathbf{r},t;\tilde{\mathbf{s}},\nu)I(\mathbf{r},t;\tilde{\mathbf{s}},\nu)]\tilde{\mathbf{s}}. \quad (2.1.8)$$

The equations (2.1.7) and (2.1.8) are the radiant energy and momentum equations in inertial or lab frame. In (2.1.7) and (2.1.8) the right hand side of the equations involve double integral terms over angle and frequency.

The direct numerical integration of transfer equation (2.1.2) with angle and frequency dependent opacity and emissivity or the moment equation (2.1.7) is possible though the process is formidable and uneconomical. The opacity and emissivity terms have to be computed taking a large number of points in angle and frequency grids. In computing χ and η angles and frequencies have to be transformed by Lorentz transformation into comoving frame. To avoid tedious and costly computations, this method is hardly used in practice.

Before attending to the over all transformation of physical quantities representing the radiation and the material field in a moving medium, we shall first consider the consequence of Doppler shifts. We shall look into the effects of local frequency transformation between the lab and comoving frame. We ignore the aberration and advection effects. For simplicity we consider a time independent radiative transfer equation in a spherically symmetric moving media in a lab frame. From (1.3.12), we have for time independent transfer equation in lab frame as

$$\frac{dI}{ds} = \mu \frac{\partial I(r,\mu,\nu)}{\partial r} + \left(\frac{1-\mu^2}{r}\right)\frac{\partial I(r,\mu,\nu)}{\partial \mu}$$
$$= \eta(r,\mu,\nu) - \chi(r,\mu,\nu)I(r,\mu,\nu). \qquad (2.1.9)$$

In (2.1.9), the space derivative $\mu(\partial/\partial r)$ is evaluated in the lab frame at constant ν. But if we move a distance Δr keeping ν constant, the frequency ν_0 [cf. (2.1.1)] will change because of the motion of the media. ν_0 is measured in the comoving frame, an imaginary frame at rest with respect to the medium at r, the point of observation. The velocity of the medium at r is $V(r)$.

The physical variables measured in the comoving frame will be distinguished by subscripts such as μ_0, ν_0, ... or superscripts as in $I^0(r,\mu_0,\nu_0)$, $\eta^0(r,\nu_0)$, $\chi^0(r,\nu_0)$, etc. We write the specific intensity in comoving frame as

$$I^0 = I^0(r,\mu_0,\nu_0), \qquad (2.1.10)$$

and

$$\frac{dI^0}{ds} = \frac{\partial I^0(r,\mu_0,\nu_0)}{\partial r}\frac{dr}{ds} + \frac{\partial I^0(r,\mu_0,\nu_0)}{\partial \mu_0}\frac{d\mu_0}{ds} + \frac{\partial I^0(r,\mu_0,\nu_0)}{\partial \nu_0}\frac{d\nu_0}{ds}.$$
$$(2.1.11)$$

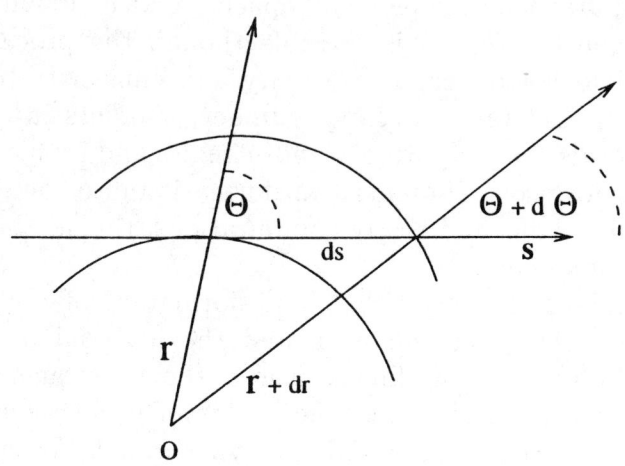

Figure 2.1. The r-θ, s coordinates for the spherical medium

From Fig (2.1),
$$\mu_0 = \cos\theta_0, \quad dr = \cos\theta_0 ds, \quad rd\theta_0 = -\sin\theta_0 ds.$$

Hence
$$\frac{dr}{ds} = \cos\theta_0 = \mu_0, \quad \frac{d\mu_0}{ds} = \frac{d\mu_0}{d\theta_0}\frac{d\theta_0}{ds} = \frac{\sin^2\theta_0}{r} = \frac{1-\mu_0^2}{r}. \quad (2.1.12)$$

Now from Doppler effect [cf.(2.1.1) with($\nu' = \nu_0$)]
$$\nu = \nu_0(1 + \frac{\mu_0 V}{c}),$$

where ν_0 is the frequency in the comoving frame and ν that in the lab frame. Hence

$$\frac{\partial \nu_0}{\partial r} = -\nu_0 \frac{\mu_0}{c} \frac{\partial V}{\partial r}, \quad \frac{\partial \nu_0}{\partial \mu_0} = \frac{\nu_0 V}{c}.$$

Therefore,

$$\begin{aligned}\frac{d\nu_0}{ds} &= \frac{\partial \nu_0}{\partial r}\frac{dr}{ds} + \frac{\partial \nu_0}{\partial \mu_0}\frac{d\mu_0}{ds} \\ &= -(\frac{\mu_0^2}{c})\nu_0 \frac{\partial V}{\partial r} - \nu_0(\frac{V}{c})\frac{1-\mu_0^2}{r}.\end{aligned} \quad (2.1.13)$$

Thus the time independent radiative transfer equation in the co-moving frame for spherically symmetric moving media can be written as

$$\mu_0 \frac{\partial I^0(r,\mu_0,\nu_0)}{\partial r} + (\frac{1-\mu_0^2}{r})\frac{\partial I^0(r,\mu_0,\nu_0)}{\partial \mu_0}$$
$$-\frac{\nu_0 V}{cr}\left[(1-\mu_0^2) + \mu_0^2(\frac{d\ln V}{d\ln r})\right]\frac{\partial I^0(r,\mu_0,\nu_0)}{\partial \nu_0}$$
$$= \eta^0(r,\nu_0) - \chi^0(r,\nu_0)I^0(r,\mu_0,\nu_0). \quad (2.1.14)$$

Some of the distinct features of the radiative transfer equation in comoving frame comes out of this restricted form.

(i) χ^0 and η^0 are isotropic in comoving frame though they were not so in the lab frame.

(ii) The change of photon frequency ν_0 as observed from lab frame is given by the frequency derivative.

(iii) The equations of type (2.1.14) lead to hyperbolic equations requiring for their solution two boundary conditions in space coordinates and an initial condition on frequency.

(iv) The scattering term in η^0 and χ^0 are evaluated in a small region around ν_0.

(v) In most physical flows $(V/c) \ll 1$. But however tempting it may be, one should not ignore terms of $O(V/c)$ even in continua without thorough analysis. On closer look it is found that it is essential to distinguish between opacity, χ and emissivity η in lab frame from χ^0 and η^0 in moving frame.

2.2 Transformation Scheme Between Variables in Comoving and Lab Frames

For proper understanding of the relation between physical variables occurring in the transfer equations in lab and comoving frames one must use an appropriate transformation scheme between the quantities in the two frames.

The transformation scheme used for this purpose is Lorentz transformation for four vectors. This corresponds to proper rotation in four dimensional space-time. The Lorentz transformation applies only between reference frames in uniform motion with respect to one another. In the presence of velocity gradient in the medium, local Lorentz transformation has to be used, when a particular velocity is assigned to every point r in spherical geometry.

We assume that a frame comoving with an element of the medium at r has a uniform velocity $V(r)$ relative to the lab frame. A point in lab frame and comoving frame is specified by four coordinates.

Lab frame:
$$(x^1, x^2, x^3, x^4) = (x, y, z, ict). \tag{2.2.1a}$$

Comoving frame:
$$(x_0^1, x_0^2, x_0^3, x_0^4) = (x_0, y_0, z_0, ict_0). \tag{2.2.1b}$$

Using the summation convention for repeated indices and assuming that the comoving frame is moving with uniform velocity V relative to the Z-axis say, we have

$$x_0^\alpha = \mathcal{L}_\beta^\alpha \, x^\beta \tag{2.2.2}$$

where

$$\mathcal{L} = \begin{pmatrix} 1 & 0 & 0 & 0 \\ 0 & 1 & 0 & 0 \\ 0 & 0 & \gamma & i\beta\gamma \\ 0 & 0 & -i\beta\gamma & \gamma \end{pmatrix} \tag{2.2.3}$$

with $\beta = V/c$ where c is the velocity of light, $i^2 = -1$, and

$$\gamma = \frac{1}{(1-\beta^2)^{1/2}}. \tag{2.2.4}$$

2.2 Transformation between Comoving and Lab frames

It is to be emphasised that four vectors and four tensors are covariant under Lorentz transformation. Hence the physical laws written in terms of four vectors or four tensors are covariant under Lorentz transformation.

Transformation of Radiative Transfer Related Physical Variables

The transformation rules for space-time are stated below, assuming that the comoving frame is moving with constant velocity V along the Z-axis. We have

(a) $\quad \Delta x_0 = \Delta x, \quad \Delta y_0 = \Delta y(x), \quad \Delta z_0 = \Delta z/\gamma,$ \hfill (2.2.5)

(b) from Lorentz–Fitzerald contraction

$$\Delta t_0 = \gamma \Delta t, \quad (2.2.6)$$

(c) from (a) and (b)

$$dV\, dt = dV_0 dt_0, \quad (2.2.7)$$

implying that space-time volume is invariant.

(d) a covariant vector four-gradient transformation as

$$\left(\frac{\partial}{\partial x}, \frac{\partial}{\partial y}, \frac{\partial}{\partial z}, \frac{1}{ic}\frac{\partial}{\partial t}\right)$$
$$= \left[\frac{\partial}{\partial x_0}, \frac{\partial}{\partial y_0}, \gamma\left(\frac{\partial}{\partial z_0} - \frac{\beta}{c}\frac{\partial}{\partial t_0}\right), \frac{\gamma}{ic}\left(\frac{\partial}{\partial t_0} - \beta c \frac{\partial}{\partial z_0}\right)\right] \quad (2.2.8)$$

We now turn to the transformation of other physical variables. Let P^α be the four momentum of any particle. Then

$$P^\alpha = \left(p_x, p_y, p_z, \frac{iE}{c}\right) \quad (2.2.9)$$

where p_x, p_y, p_z are the components of the ordinary three momentum and E, the total energy of the particle. We know

$$p^2 = p_x^2 + p_y^2 + p_z^2, \quad (2.2.10)$$

and
$$E^2 = p^2 c^2 + (m_0 c^2)^2, \qquad (2.2.11)$$
where m_0 is the rest mass of the particle.

For photons,
$$\text{the rest mass } m_0 = 0, \quad E = h\nu, \quad p = h\nu/c, \qquad (2.2.12)$$
where h is the Planck constant.
$$P^\alpha = \frac{h\nu}{c}(n_x^0, n_y^0, n_z^0, i). \qquad (2.2.13)$$

Then the unit vector in the direction of photon propagation is
$$\mathbf{n} = (n_x^0, n_y^0, n_z^0) ;$$
$$n_x^0 = \sin\theta_0 \cos\varphi_0, \quad n_y^0 = \sin\theta_0 \sin\varphi_0, \quad n_z^0 = \cos\theta_0.$$

Now we apply Lorentz transformation to (2.2.13) assuming that the comoving frame is moving with constant velocity V along the Z-axis of the lab frame. We have
$$[\nu_0 n_x^0, \nu_0 n_y^0, \nu_0 n_z^0, i\nu_0] = [\nu n_x, \nu n_y, \gamma\nu(n_z - \beta), i\nu\gamma(1 - n_z\beta)]. \qquad (2.2.14)$$

This implies
$$\nu_0 = \gamma\nu(1 - n_z\beta) = \gamma\nu(1 - \mu\beta). \qquad (2.2.15)$$

In view of (2.2.15), in the axial symmetric case, we can write
$$[(1 - \mu_0^2)^{1/2}, \mu_0, \nu_0] = \left[\frac{\gamma^{-1}(1 - \mu^2)}{(1 - \mu\beta)}, \frac{(\mu - \beta)}{(1 - \mu\beta)}, \gamma\nu(1 - \mu\beta)\right]. \qquad (2.2.16)$$

Equation (2.2.16) is equivalent to (2.2.14).

The inverse transformation (replacing β by $-\beta$) is
$$[(1 - \mu^2)^{1/2}, \mu, \nu] = \left[\frac{\gamma^{-1}(1 - \mu_0^2)}{(1 + \mu_0\beta)}, \frac{(\mu_0 + \beta)}{(1 + \mu_0\beta)}, \gamma\nu_0(1 + \mu_0\beta)\right]. \qquad (2.2.17)$$

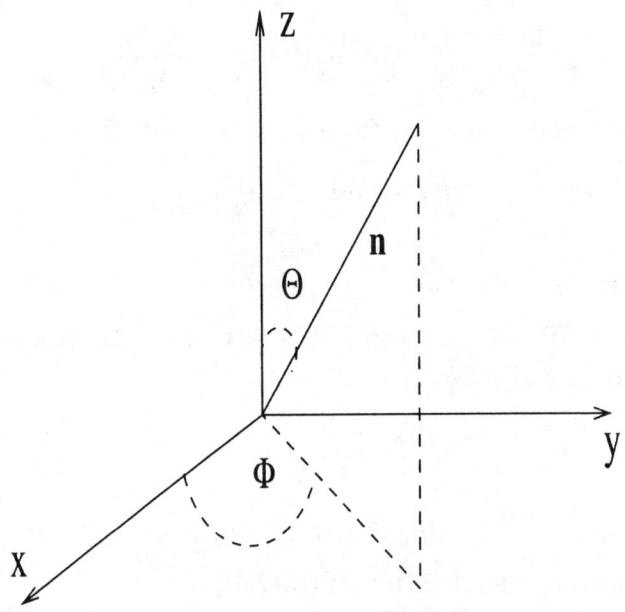

Figure 2.2. Direction of ray in Cartesian coordinate system

From (2.2.16), we have

$$\nu_0 = \gamma\nu(1 - \mu\beta), \quad \text{Doppler shift,} \qquad (2.2.18)$$

$$\mu_0 = \frac{\mu - \beta}{1 - \mu\beta}, \quad \text{aberration effect.} \qquad (2.2.19)$$

From (2.2.17) we can get the inverse transformations. Now retaining only terms upto $O(V/c)$ and putting $\gamma = 1$, we obtain the expressions for Doppler shift and aberration in the classical Galilean approximation. They are

$$\nu_0 = \nu(1 - \mu\beta), \qquad (2.2.20)$$

$$\mu_0 = \frac{\mu - \beta}{1 - \mu\beta}, \qquad (2.2.21)$$

$$d\nu_0 = (1 - \mu\beta)d\nu, \qquad (2.2.22)$$

$$d\mu_0 = \frac{d\mu}{1 - \mu\beta} + \frac{(\mu - \beta)\beta}{(1 - \mu\beta)^2}d\mu, \approx \frac{d\mu}{(1 - \mu\beta)^2}. \qquad (2.2.23)$$

In the non-classical case, from (2.2.18) and (2.2.19)

$$d\nu_0 = \gamma(1 - \mu\beta)d\nu, \qquad (2.2.24)$$

$$d\mu_0 = \frac{d\mu}{(1 - \mu\beta)^2}. \qquad (2.2.25)$$

From (2.2.20), (2.2.22) and (2.2.23) and also from (2.2.18), (2.2.24) and (2.2.25), we have

$$d\nu_0 = (\nu_0/\nu)d\nu, \qquad (2.2.26)$$

and

$$d\mu_0 = (\nu/\nu_0)^2 d\mu, \qquad (2.2.27)$$

where we have retained terms upto $O(V/c)$.

We now take note of the fact that an elementary solid angle in spherical coordinates

$$d\Omega = \sin\theta d\theta d\varphi = |\, d\mu d\varphi \,|. \qquad (2.2.28)$$

Then from (2.2.18), (2.2.26)–(2.2.28) we get

$$\nu d\nu d\Omega = \nu_0 d\nu_0 d\Omega_0. \qquad (2.2.29)$$

This implies that $\nu d\nu d\Omega$ is Lorentz invariant.

Transformation of the Specific Intensity of Radiation

From the definition of the specific intensity [cf. (1.1.1)] the number of photons passing through an elementary area $d\Sigma$ (stationary) placed at right angles to the Z-axis in frequency interval $d\nu$ into a solid angle $d\Omega$ moving at an angle $\cos^{-1}\mu$ to the Z-axis is given by

$$N = \left[\frac{I(r, t; \mu, \nu)}{h\nu} d\Omega d\nu d\Sigma \cos\theta dt\right], \qquad (2.2.30)$$

to an observer in the lab frame.

An observer in the comoving frame will count the number as

$$N_0 = \left[\frac{I^0(r,t;\mu_0,\nu_0)}{h\nu_0}d\Omega_0 d\nu_0 \left(d\Sigma\cos\theta_0 dt_0 + \frac{d\Sigma V dt_0}{c}\right)\right]. \tag{2.2.31}$$

The first term on the right hand side gives the number of photons that would have been counted in case $d\Sigma$ was stationary in the comoving frame. The second term accounts for the number of photons present in the volume swept over and is equal to the product of the density of photons $I^0/(ch\nu_0)$ and the volume swept over $d\Sigma V dt_0$.

Then from (2.2.5)–(2.2.7), (2.2.18), (2.2.19) (2.2.30) and (2.2.31) we have

$$\left[\left(\frac{I(r,t;\mu,\nu)}{\nu}\right)d\Omega d\nu d\Sigma \mu dt\right]$$
$$= \left[\left(\frac{I^0(r,t;\mu_0,\nu_0)}{\nu_0}\right)d\Omega_0 d\nu_0 d\Sigma dt_0\right](\mu_0 + \beta). \tag{2.2.32}$$

Multiplying both sides of (2.2.32) by $\nu^2 \nu_0^2$, using (2.2.6),(2.2.29) and (2.5.6) we have,

$$I(r,t;\mu,\nu) = \left(\frac{\nu}{\nu_0}\right)^3 I^0(r,t;\mu_0,\nu_0). \tag{2.2.33}$$

Transformation of Emissivity

The number of photons emitted from a given volume into a given solid angle and frequency interval in a definite time will be the same in lab and in comoving frame. Hence if $\eta(r,t;\mu,\nu)$ is the emissivity,

$$\frac{[\eta(r,t;\mu,\nu)d\Omega d\nu dV dt]}{h\nu} = \frac{[\eta^0(r,t;\nu_0)d\Omega_0 d\nu_0 dV_0 dt_0]}{h\nu_0}. \tag{2.2.34}$$

Then from (2.2.7), (2.2.29), we have

$$\eta(r,t;\mu,\nu) = \left(\frac{\nu}{\nu_0}\right)^2 \eta^0(r,t;\nu_0). \tag{2.2.35}$$

In (2.2.35), we note that η^0 is isotropic in the comoving frame.

Transformation of Opacity or Extinction Coefficient

Similar to the case of emissivity, the number of photons absorbed in a specific element of material at a definite frequency interval and a solid angle in a given time interval will be the same as counted from a lab frame or comoving frame. Hence

$$\frac{[\chi(r,t;\mu,\nu)I(r,t;\mu,\nu)d\Omega d\nu dV dt]}{h\nu}$$
$$= \frac{[\chi^0(r,t;\mu_0,\nu_0)I^0(r,t;\mu_0,\nu_0)d\Omega_0 d\nu_0 dV_0 dt_0]}{h\nu_0}. \qquad (2.2.36)$$

Invoking the same relations as in the case of emissivity and (2.2.33), we have

$$\chi(r,t;\mu,\nu) = (\frac{\nu_0}{\nu})\chi^0(r,t;\nu_0), \qquad (2.2.37)$$

where χ^0 is again isotropic in the comoving frame.

2.3 Transformation of Radiative Transfer Equation

For transforming radiative transfer equation, we take note of the fact that Lorentz transformation of a four gradient is given by [cf. (2.2.8)]

$$\left[\frac{\partial}{\partial x}, \frac{\partial}{\partial y}, \frac{\partial}{\partial z}, \frac{1}{ic}\frac{\partial}{\partial t}\right]$$
$$= \left[\frac{\partial}{\partial x_0}, \frac{\partial}{\partial y_0}, \gamma(\frac{\partial}{\partial z_0} - \frac{\beta}{c}\frac{\partial}{\partial t_0}), \frac{\gamma}{ic}(\frac{\partial}{\partial t_0} - c\beta\frac{\partial}{\partial z_0})\right], \qquad (2.3.1)$$

$$(n_x, n_y, n_z) = \frac{\nu_0}{\nu}\left[n_x^0, n_y^0, (1/\gamma)(n_z^0 + \beta)\right], \qquad (2.3.2)$$

and

$$\frac{1}{c}\frac{\partial}{\partial t} = \frac{\gamma}{c}(\frac{\partial}{\partial t_0} - c\beta\frac{\partial}{\partial z_0}). \qquad (2.3.3)$$

2.3 Transformation of Radiative Transfer Equation

Hence

$$\left[\frac{1}{c}(\frac{\partial}{\partial t}) + \mathbf{n} \cdot \nabla\right] = (\frac{\nu_0}{\nu})\left[\frac{1}{c}(\frac{\partial}{\partial t_0}) + \mathbf{n}^0 \cdot \nabla_0\right]. \qquad (2.3.4)$$

Then the transfer equation

$$\frac{1}{c}\frac{\partial I(r,t;\mu,\nu)}{\partial t} + (\mathbf{n} \cdot \nabla)I(r,t;\mu,\nu)$$
$$= \eta(r,t;\mu,\nu) - \chi(r,t;\mu,\nu)I(r,t;\mu,\nu), \qquad (2.3.5)$$

transforms to [cf. (2.2.33), (2.2.35), (2.2.37) and (2.3.4)]

$$(\frac{\nu_0}{\nu})\left[\frac{1}{c}\frac{\partial}{\partial t_0} + \mathbf{n}^0 \cdot \nabla_0\right](\frac{\nu}{\nu_0})^3 I^0(r,t;\mu_0,\nu_0)$$
$$= (\frac{\nu}{\nu_0})^2\left[\eta^0(r,t;\nu_0) - \chi^0(r,t;\nu_0)I^0(r,t;\mu_0,\nu_0)\right],$$

which implies

$$\left[\frac{1}{c}(\frac{\partial}{\partial t_0}) + \mathbf{n}^0 \cdot \nabla_0\right]I^0(r,t;\mu_0,\nu_0)$$
$$= \left[\eta^0(r,t;\nu_0) - \chi^0(r,t;\nu_0)I^0(r,t;\mu_0,\nu_0)\right]. \qquad (2.3.6)$$

The forms of equations (2.3.5) and (2.3.6) suggest that the radiative transfer equations in the two frames of reference moving in uniform motion with respect to one another are covariant. Thus the radiative transfer equation in lab frame at rest is covariant with a local comoving frame planted at a particular point r in the medium. However, the covariance of the transfer equations in lab frame and comoving frame at all points of an expanding or pulsating atmospheres cannot obviously be assumed as the velocity $V(r)$ in these cases is a function of r.

Transformation of the Moments of the Radiation Intensity

In equations (1.1.8)–(1.1.33), we have introduced the monochromatic and frequency integrated angular moments of the specific intensities. We defined mean intensity J, radiative flux $F(H = F/4)$ and K-integral of the radiation field. We have also established the relations between $J, H(= F/4), K$ with energy density E of radiation, energy flux F and radiation pressure tensor or radiation stress tensor $\mathcal{P}$. The moment equations for radiative transfer for monochromatic and frequency integrated moments of radiation intensity have been deduced in equations (2.1.4)–(2.1.8) and they have been expressed in terms of variations of energy density, flux and pressure. The moment equations were set in lab frame.

We now transform E, F, $\mathcal{P}$ expressed in lab frame or their equivalents $J, H,$ and K into comoving frame retaining upto terms of $O(V/c)$.

We recall from (2.2.16), (2.2.20), (2.2.29) and (2.2.33)

$$\nu d\nu d\Omega = \nu_0 d\nu_0 d\Omega_0,$$

$$I(r,t;\mu,\nu) = \left(\frac{\nu}{\nu_0}\right)^3 I_0(r,t;\mu_0,\nu_0),$$

$$\nu_0 = \nu(1-\mu\beta),$$

$$[(1-\mu_0^2)^{1/2}, \mu_0, \nu_0] = \left[\frac{\gamma^{-1}(1-\mu^2)^{1/2}}{(1-\mu\beta)}, \frac{(\mu-\beta)}{(1-\mu\beta)}, \gamma\nu(1-\mu\beta)\right].$$

Setting $\gamma = 1$ and expanding to first order in (V/c), we have

$$I^0 d\nu_0 d\Omega_0 = \left(\frac{\nu_0}{\nu}\right)^2 I d\nu d\Omega = I(1-2\mu\beta) d\nu d\Omega. \qquad (2.3.7)$$

This implies that

$$J_0 = J - 2\beta H. \qquad (2.3.8)$$

Similarly we may write

$$\begin{aligned}
I^0 \mu_0 d\nu_0 d\Omega_0 &= I\left(\frac{\mu-\beta}{1-\mu\beta}\right)(1-\mu\beta)^2 d\nu d\Omega, \\
&= I(\mu-\beta)(1-\mu\beta) d\nu d\Omega \\
&\approx \left[\mu - \beta(1+\mu^2)\right] I d\nu d\Omega, \qquad (2.3.9)
\end{aligned}$$

$$I_0\mu_0^2 d\nu_0 d\Omega_0 = (\mu - \beta)^2 I d\nu d\Omega,$$
$$= (\mu^2 - 2\beta\mu)I d\nu d\Omega. \quad (2.3.10)$$

From (2.3.9) and (2.3.10), it follows that

$$H^0 = H - \beta(J + K), \quad (2.3.11)$$

and

$$K^0 = K - 2\beta H. \quad (2.3.12)$$

J^0, H^0, K^0, and J, H, K are frequency integrated angular moments in comoving and lab frame respectively.

The inverse of these transformations give (replacing β by $-\beta$)

$$[J, H, K] = \left[J^0 + 2\beta H^0, H^0 + \beta(J^0 + K^0), K^0 + 2\beta H^0\right]. \quad (2.3.13)$$

2.4 Mixed Equations

In equation (2.1.3), we have stated the radiative transfer equation in spherically symmetric moving media in lab frame. In (2.1.5) and (2.1.6), we have deduced the moment equations of transfer and expressed them in terms of variation of monochromatic energy density, flux and pressure tensor. The opacity χ and emissivity η in all the above cases are angle and frequency dependent and anisotropic.

In the mixed approach, the space and time derivatives on the left hand side of the transfer equation and the physical variables representing the radiation field, angles and frequencies are measured in lab frame. (V/c) is assumed much less than unity $((V/c) \ll 1)$. The opacity χ and the emissivity η are evaluated through first order expansions at the appropriate comoving frame of the atmospheric medium.

Then remembering the relations [cf. (2.2.20),(2.2.35), (2.2.37)]

$$\nu = \nu_0(1 + \mu\beta) = \nu_0(1 + \tilde{\mathbf{s}} \cdot \mathbf{V}/c), \quad (2.2.20)$$

$$\eta(\mathbf{r},t;\mathbf{n},\nu) = (\frac{\nu}{\nu_0})^2 \eta^0(\mathbf{r},t;\mathbf{n}_0,\nu_0), \qquad (2.2.35)$$

$$\chi(\mathbf{r},t;\mathbf{n},\nu) = (\frac{\nu_0}{\nu}) \chi^0(\mathbf{r},t;\mathbf{n}_0,\nu_0), \qquad (2.2.37)$$

we have from (2.2.37)

$$\chi(\mathbf{n},\nu) = \chi^0(\nu) - \frac{(\mathbf{n}\cdot\mathbf{V})}{c}\left[\chi^0(\nu) + \nu(\frac{\partial\chi^0}{\partial\nu})\right] \qquad (2.4.1)$$

and from (2.2.35)

$$\eta(\mathbf{n},\nu) = \eta^0(\nu) + \frac{(\mathbf{n}\cdot\mathbf{V})}{c}\left[2\eta^0(\nu) - \nu(\frac{\partial\eta^0}{\partial\nu})\right]. \qquad (2.4.2)$$

Hence we can write the transfer equation in the mixed form as

$$\left[\frac{1}{c}(\frac{\partial}{\partial t}) + \mathbf{n}\cdot\boldsymbol{\nabla}\right]I(r,t;\mathbf{n},\nu)$$
$$= \eta^0(\nu) - \chi^0(\nu)I(r,t;\mathbf{n},\nu) + \frac{\mathbf{n}\cdot\mathbf{V}}{c}\left[2\eta^0(\nu) - \nu\frac{\partial\eta^0}{\partial\nu}\right.$$
$$\left.+\{\chi^0(\nu) + \nu\frac{\partial\chi^0}{\partial\nu}\}I(r,t;\mathbf{n},\nu)\right]. \qquad (2.4.3)$$

In equation (2.4.3) χ^0 and η^0 are in comoving frame and hence are isotropic. For calculating intensity in continua the mixed form of equation is expected to yield fairly reliable results, because the opacity and emissivity changes very little over a Doppler width. However the scheme fails in the case of spectral line, since in this case the first order expansion in $\Delta\nu$ is not adequate to account for rapid variation of opacity and emissivity over most of the line profiles.

Integrating (2.4.3) over angle and frequency, we obtain the radiation energy and momentum equations as zeroth and first angular moment equations respectively.

The energy equation is

$$\frac{\partial E}{\partial t} + \boldsymbol{\nabla}\cdot\mathbf{F} = \int_0^\infty [4\pi\eta^0(\nu) - \chi^0(\nu)cE_\nu]d\nu$$
$$+(\frac{\mathbf{V}}{c})\cdot\int_0^\infty \mathbf{F}_\nu\left[\chi^0(\nu) + \nu\frac{\partial\chi^0}{\partial\nu}\right]d\nu \qquad (2.4.4)$$

and the radiation momentum equation

$$\frac{1}{c^2}\frac{\partial \mathbf{F}}{\partial t} + \nabla \cdot \mathcal{P} = -\frac{1}{c}\int_0^\infty \chi^0(\nu)\mathbf{F}d\nu + \frac{4\pi \mathbf{V}}{c^2}\int_0^\infty \eta^0(\nu)d\nu$$
$$+(\frac{\mathbf{V}}{c})\cdot\int_0^\infty \left[\chi^0(\nu) + \nu\frac{\partial \chi^0(\nu)}{\partial \nu}\right]\mathcal{P}_\nu d\nu. \quad (2.4.5)$$

We recall here that in (2.4.4) and (2.4.5) χ^0 and η^0 are measured in comoving frame whereas the space-time variables, radiation field variables, angles and frequencies are measured in lab frame.

Mihalas and Mihalas (1984, p. 423–426) made thorough inquiry into the necessity of retaining the terms of $O(V/c)$ in the mixed form of equations on the right hand side. We summarise below the salient points of their argument in favour of retaining terms $O(V/c)$ in some cases. For simplicity, the moment equations in grey approximation were considered. The energy and momentum equation in that case become

$$\frac{\partial E}{\partial t} + \nabla \cdot \mathbf{F} = \chi^0(4\pi B - cE) + \frac{\chi^0}{c}\mathbf{V}\cdot\mathbf{F}, \quad (2.4.6)$$

and

$$\frac{1}{c^2}\frac{\partial \mathbf{F}}{\partial t} + \nabla \cdot \mathcal{P} = -\frac{\chi^0}{c}\mathbf{F} + \frac{\chi^0}{c}\left[\frac{4\pi B}{c}\mathbf{V} + \mathbf{V}\cdot\mathcal{P}\right]. \quad (2.4.7)$$

(A) First we consider the energy equation (2.4.6). The velocity dependent term on the right hand side measures the rate of work done by the radiation force on the material. When this force is strong, the omission of this term may give rather costly error. In the diffusion limit $(\lambda_p/l) \ll 1$, where (λ_p is the photon mean free path and l is the characteristic length) $E_R^0 \to (4\pi B/c)$.
We recall that [cf. (1.1.17), (1.1.21), (1.1.23), (1.1.31), (2.3.13)]

$$[J, H, K] = \left[J^0 + 2\beta H^0, H^0 + \beta(J^0 + K^0), K^0 + 2\beta H^0\right]$$

and
$$E_R = \frac{4\pi J_R}{c}, \quad H_R = \frac{\bar{F}_R}{4}, \quad P_R = \frac{E_R}{3}$$
for isotropic radiation field. The suffix R indicates that the quantities mentioned are radiative components. These imply

$$\begin{aligned} E_R &= E_R^0 + \frac{2\beta \bar{F}_R^0}{c}, \\ \bar{F}_R &= \bar{F}_R^0 + VE_R^0 + VP_R^0, \\ P_R &= P_R^0 + \frac{2\beta \bar{F}_R^0}{c}. \end{aligned} \quad (2.4.8)$$

We shall suppress the suffixes in the rest of this section. Hence

$$4\pi B - cE = -2\beta F = -\frac{2\mathbf{V} \cdot \mathbf{F}}{c} + O(V^2/c^2). \quad (2.4.9)$$

Hence the energy equation (2.4.6) becomes

$$\frac{\partial E}{\partial t} + \nabla \cdot \mathbf{F} = -\frac{\chi^0}{c}\mathbf{V} \cdot \mathbf{F}. \quad (2.4.10)$$

This implies that the rate of change of radiation energy density equals the flow of radiation out of the enclosed region less the rate of work done by the radiation field on the material. The following two points are to be noted.

(i) The omission of (V/c) term on the right hand side amounts to ignoring absorption and emission in the diffusion limit.

(ii) In the dynamic diffusion region where $(V/c) \geq (\lambda_p/l)$ the ratio of $[\chi \mathbf{V} \cdot \mathbf{F}/c]$ to $\nabla \cdot \mathbf{F}$ is of the order of $O(Vl/c\lambda_p)$ where λ_p is the photon mean free path. So it is reasonable to assume that in this regime velocity dependent term is the dominant factor.

(B) Now we consider the radiation momentum equation (2.4.7). To study this, it is instructive to distinguish between the role of different terms in two limiting cases of (a) **streaming** regions which is relevant in optically thin media and (b) **diffusion limit** which is relevant to stellar interior or optically thick region.

In streaming limit, two characteristic time scales are analysed. They are the radiation flow time scale, $t_R = l/c$ and fluid flow time scale, $t_f = l/V$. Here l is the characteristic length of the material and $(\lambda_p/l) \geq 1$.

In the diffusion limit, we distinguish between static diffusion when $t_f \gg t_d$, where t_d is the radiation diffusion time given by $t_d = l^2/c\lambda_p$ and $(V/c) \ll (\lambda_p/l)$ and the dynamic diffusion when $t_f \ll t_d$, and $(V/c) \gg (\lambda_p/l)$.

On a fluid flow time scale, the time derivative in the momentum equation (2.4.7) is very small and is of $O(\lambda_p V/lc)$. So in the diffusion regime where the medium is optically thick, the time derivative can be neglected and (2.4.7) can be written as

$$\mathbf{F} = -\frac{c}{\chi}\nabla \cdot \mathcal{P} + (\frac{4\pi B}{c})\mathbf{V} + \mathbf{V} \cdot \mathcal{P}. \tag{2.4.11}$$

In the diffusion regime, considering terms of $O(V/c)$, we have

$$E^0 \to (\frac{4\pi B}{c}),$$

$$P^{ij} \to P^{0ij} + O(\frac{\lambda_p V}{lc}) = \frac{E^0 \delta^{ij}}{3} + O(\frac{\lambda_p V}{lc})$$

$$\mathbf{F}^0 \to -(\frac{c}{\chi})\nabla \cdot \mathcal{P}. \tag{2.4.12}$$

From (2.4.11) and (2.4.12)

$$\begin{aligned} F &= F^0 + E^0 \mathbf{V} + \mathbf{V}.\mathcal{P}^0 + O(\frac{V^2}{c^2}) \\ &= F^0 + (\frac{4}{3})E^0 \mathbf{V}. \end{aligned} \tag{2.4.13}$$

This is precisely the Lorentz transformation of F^0 from Eulerian lab frame to Lagrangian comoving frame. We thus conclude that exclusion of terms $O(V/c)$ in the momentum equation (2.4.7) will not allow us to distinguish between fluxes in the lab frame and comoving frame.

From the above, we may conclude that term by term analysis of the importance of each term in the transfer and momentum equations must be made before dropping terms $O(V/c)$ from the right hand side of the transfer and momentum equations. In general, flows cover both optically thin as well as optically thick regions involving different magnitudes of velocities. A term which makes inconsequential contribution in one region may play a dominant role in the other regions.

We state below some of the guide lines about the relative importance of the different terms in solving radiative transfer equations of mixed form. For more detailed analysis and discussions on this matter, the readers are referred to Mihalas and Mihalas (1984, p. 423–426).

(a) To solve the mixed frame transfer equation (2.4.3) satisfactorily on a fluid time scale t_f, the spatial differential operator on the left hand side of the equation and all the terms on the right hand side are retained. The $(\partial/\partial t)$ term can be dropped.

(b) The moment equations (2.4.4) and (2.4.5) for radiation energy and momentum can be solved correctly on the time scale t_f, when all the terms of the two equations retained except for the $(\partial/\partial t)$ term in the momentum equation.

(c) In solving the transfer equation (2.4.3) and the moment equation (2.4.4) on a radiative time scale t_R, all the terms including the $(\partial/\partial t)$ term have to be retained.

By following the above guide lines, one can be sure that the dominant terms in all the regions have been taken care of properly.

2.5 Radiative Transfer Equations in Spherically Symmetric Moving Media in Comoving Frame

The comoving frame of a fluid volume may be supposed to consist of sets of inertial frames attached to the elements of volume each of which is identified by a velocity instantaneously possessed by each element in the volume. The change of physical variables between the Eulerian lab frame and the Lagrangian comoving frame is obtained by Lorentz transformation, which strictly speaking, applies to frames in uniform motion relative to one another. However, in stellar atmospheres velocity of the medium $V = V(r,t)$ is time dependent, making the fluid frames non-inertial. The method employed for setting up the comoving frame transfer equation is the

2.5 Transfer Equations in CMF

use of an infinite set of local inertial frames. This allows us to take advantage of the Lorentz transformation of the physical variables of radiation field and matter between the lab frame and the local comoving frames.

We have two options at our hand for establishing the transfer equations in comoving frame.

(a) We may use the inertial space time structure and use special relativity to transform the variables and derive the equation retaining terms upto $O(V/c)$ or

(b) we may take note of the non-inertial nature of the frames involved in moving stellar atmospheres and use the techniques of general relativity to deduce the equation. We shall mainly consider the procedure (a) but the reference to (b) will be included.

For simplicity, we shall confine ourselves to one dimensional spherically symmetric flows. The transfer equation in lab frame for a spherically symmetric radiation flow is given by [cf. (1.3.7)]

$$\left[\frac{1}{c}\frac{\partial}{\partial t} + \mu\frac{\partial}{\partial r} + \frac{1-\mu^2}{r}\frac{\partial}{\partial \mu}\right] I(r,t;\mu,\nu)$$
$$= \eta(r,t;\mu,\nu) - \chi(r,t;\mu,\nu)I(r,t;\mu,\nu). \qquad (2.5.1)$$

From (2.2.33), (2.2.35) and (2.2.37), we have

$$I(\mu,\nu) = \left(\frac{\nu}{\nu_0}\right)^3 I^0(\mu_0,\nu_0)$$

$$\eta(\mu,\nu) = \left(\frac{\nu}{\nu_0}\right)^2 \eta^0(\nu_0)$$

$$\chi(\mu,\nu) = \left(\frac{\nu_0}{\nu}\right)\chi^0(\nu_0).$$

In the above, we have suppressed the variables (r,t) in representing I, η, χ and their transformations for simplifying the notation. We shall continue to do so in the rest of this chapter.

With the above substitutions in (2.5.1), we have

$$\left(\frac{\nu}{\nu_0}\right)\left[\frac{1}{c}\frac{\partial}{\partial t} + \mu\frac{\partial}{\partial r} + \left(\frac{1-\mu^2}{r}\right)\frac{\partial}{\partial \mu}\right] I^0(\mu_0,\nu_0)$$

$$-3\left(\frac{\nu}{\nu_0^2}\right)\left[\frac{1}{c}\frac{\partial \nu_0}{\partial t} + \mu\frac{\partial \nu_0}{\partial r} + \left(\frac{1-\mu^2}{r}\right)\frac{\partial \nu_0}{\partial \mu}\right]I^0(\mu_0,\nu_0)$$
$$= \eta^0(\nu_0) - \chi^0(\nu_0)I^0(\mu_0,\nu_0). \qquad (2.5.2)$$

The quantities with suffix and affix are measured in the Lagrangian comoving frame. The derivatives $\partial/\partial r$ and $\partial/\partial t$ are calculated on the assumption that μ and ν are held constant. However, this restriction is not imposed in the case of $\partial/\partial \mu$ variations. We should also note that the velocity $V(r,t)$ of the medium changes with space and time, μ_0 and ν_0 are not constants. Hence care must be taken in the variations of μ_0 and ν_0. We apply the chain rule to determine the derivatives.

$$\left(\frac{\partial}{\partial t}\right)_{r\mu\nu} = \left(\frac{\partial}{\partial t}\right)_{r\mu_0\nu_0} + \left(\frac{\partial \mu_0}{\partial t}\right)_{r\mu\nu}\frac{\partial}{\partial \mu_0}$$
$$+ \left(\frac{\partial \nu_0}{\partial t}\right)_{r\mu\nu}\frac{\partial}{\partial \nu_0} + \left(\frac{\partial t_0}{\partial t}\right)_{r\mu\nu}\frac{\partial}{\partial t_0} \qquad (2.5.3)$$

$$\left(\frac{\partial}{\partial r}\right)_{t\mu\nu} = \left(\frac{\partial}{\partial r}\right)_{t\mu_0\nu_0} + \left(\frac{\partial \mu_0}{\partial r}\right)_{t\mu\nu}\frac{\partial}{\partial \mu_0}$$
$$+ \left(\frac{\partial \nu_0}{\partial r}\right)_{t\mu\nu}\frac{\partial}{\partial \nu_0} + \left(\frac{\partial t_0}{\partial r}\right)_{t\mu\nu}\frac{\partial}{\partial t_0} \qquad (2.5.4)$$

$$\left(\frac{\partial}{\partial \mu}\right)_{rt\nu} = \left(\frac{\partial \mu_0}{\partial \mu}\right)_{rt\nu}\frac{\partial}{\partial \mu_0} + \left(\frac{\partial \nu_0}{\partial \mu}\right)_{rt\nu}\frac{\partial}{\partial \nu_0} + \left(\frac{\partial t_0}{\partial \mu}\right)_{rt\nu}\frac{\partial}{\partial t_0}. \qquad (2.5.5)$$

In this if we use the expressions in first order of $\beta = (V/c)$, we have for example

$$\nu = \nu_0(1+\beta\mu_0), \quad \nu_0 = \nu(1-\beta\mu),$$
$$\mu_0 = \frac{\mu-\beta}{1-\beta\mu}, \quad \mu = \frac{\mu_0+\beta}{1+\beta\mu_0}. \qquad (2.5.6)$$

Further if we assume that

(a) the change of velocity within the flight of photon mean free path is negligible compared to the velocity,

(b) the terms like $(\partial r_0/\partial t), (\partial \mu_0/\partial t), (\partial \nu_0/\partial t)$ can be ignored, and

(c) $(\partial t_0/\partial t) = 1$,

then to terms of order (V/c),

$$\frac{\partial r_0}{\partial r} = 1, \quad \frac{\partial \mu_0}{\partial r} = (\frac{\mu_0{}^2 - 1}{c})\frac{\partial V}{\partial r_0} \qquad (2.5.7)$$

$$\frac{\partial \nu_0}{\partial r} = -\frac{\mu_0 V}{c}\frac{\partial V}{\partial r_0}, \quad \frac{\partial t_0}{\partial r} = -\frac{\beta}{c} \qquad (2.5.8)$$

$$\frac{\partial r_0}{\partial \mu} = 0, \quad \frac{\partial \mu_0}{\partial \mu} = (1 + 2\mu_0)\beta, \quad \frac{\partial \nu_0}{\partial \mu} = -\nu_0 \beta, \quad \frac{\partial t_0}{\partial \mu} = 0. \quad (2.5.9)$$

Substituting (2.5.6) and (2.5.7)–(2.5.9) in (2.5.2) we obtain the transfer equation in comoving frame to $O(V/c)$ as

$$\left[\frac{1}{c}\frac{\partial}{\partial t_0} + (\mu_0 + \frac{V}{c})\frac{\partial}{\partial r} + \frac{1-\mu_0^2}{r}\left\{1 + \frac{\mu_0 V}{c}(1 - \frac{d\ln V}{d\ln r})\right\}\frac{\partial}{\partial \mu_0}\right.$$
$$- (\frac{\nu_0 V}{cr})\left\{1 - \mu_0^2(1 - \frac{d\ln V}{d\ln r})\right\}\frac{\partial}{\partial \nu_0}$$
$$\left. + (\frac{3V}{cr})\left\{1 - \mu_0^2(1 - \frac{d\ln V}{d\ln r})\right\}\right] I^0(r_0, t; \mu_0, \nu_0)$$
$$= \eta^0(\nu_0) - \chi^0(\nu_0) I^0(r_0, t; \mu_0, \nu_0). \qquad (2.5.10)$$

Now if the equations in (2.5.6) are replaced by their special relativistic equations, we get

$$(\frac{\nu_0}{\nu}) = \gamma(1 - \beta\mu), \quad \frac{\nu}{\nu_0} = \gamma(1 + \beta\mu_0),$$

$$\mu_0 = \frac{\mu - \beta}{1 - \beta\mu}, \quad \mu = \frac{\mu_0 + \beta}{1 + \beta\mu_0}. \qquad (2.5.11)$$

where

$$\gamma = (1 - \beta^2)^{-1/2}, \quad \text{and} \quad \beta = \frac{V}{c}.$$

Hence we get

$$\left(\frac{\partial \mu_0}{\partial t}\right) = -\gamma^2(1-\mu_0^2)\left(\frac{\partial \beta}{\partial t}\right) \qquad (2.5.12)$$

$$\left(\frac{\partial \nu_0}{\partial t}\right) = -\gamma^2 \mu_0 \nu_0 \left(\frac{\partial \beta}{\partial t}\right) \qquad (2.5.13)$$

$$\left(\frac{\partial \mu_0}{\partial r}\right) = -\gamma^2(1-\mu_0^2)\left(\frac{\partial \beta}{\partial r}\right) \qquad (2.5.14)$$

$$\left(\frac{\partial \nu_0}{\partial r}\right) = -\gamma^2 \mu_0 \nu_0 \left(\frac{\partial \beta}{\partial r}\right) \qquad (2.5.15)$$

$$\left(\frac{\partial \mu_0}{\partial \mu}\right) = \gamma^2(1+\beta\mu_0)^2 \qquad (2.5.16)$$

$$\left(\frac{\partial \nu_0}{\partial \mu}\right) = -\beta\gamma^2(1+\beta\mu_0)\nu_0. \qquad (2.5.17)$$

Then using (2.5.3)–(2.5.5), (2.5.11)–(2.5.17), and substituting in (2.5.2), we have for the radiative transfer equation in comoving frame as

$$\begin{aligned}
&\frac{\gamma}{c}(1+\beta\mu_0)\frac{\partial I^0(\mu_0,\nu_0)}{\partial t} + \gamma(\mu_0+\beta)\frac{\partial I^0(\mu_0,\nu_0)}{\partial r} \\
&+ \frac{\partial}{\partial \mu_0}\left[\gamma(1-\mu_0^2)\left\{\frac{(1+\beta\mu_0)}{r} - \gamma^2(\mu_0+\beta)\frac{\partial \beta}{\partial r}\right.\right. \\
&\left.\left. - \frac{\gamma^2}{c}(1+\beta\mu_0)\frac{\partial \beta}{\partial t}\right\}I^0(\mu_0,\nu_0)\right] \\
&- \frac{\partial}{\partial \nu_0}\left[\gamma\nu_0\left\{\frac{\beta(1-\mu_0^2)}{r} + \gamma^2\mu_0(\mu_0+\beta)\frac{\partial \beta}{\partial r}\right.\right. \\
&\left.\left. + \frac{\gamma^2}{c}\mu_0(1+\beta\mu_0)\frac{\partial \beta}{\partial t}\right\}I^0(\mu_0,\nu_0)\right] \\
&+ \gamma\left[\frac{2\mu_0+\beta(3-\mu_0^2)}{r} + \gamma^2(1+\mu_0^2+2\beta\mu_0)\frac{\partial \beta}{\partial r}\right. \\
&\left. + \frac{\gamma^2}{c}\{2\mu_0+\beta(1+\mu_0^2)\}\frac{\partial \beta}{\partial t}\right]I^0(\mu_0,\nu_0) \\
&= \eta^0(\nu_0) - \chi^0(\nu_0)I^0(\mu_0,\nu_0). \qquad (2.5.18)
\end{aligned}$$

The equation (2.5.18) is the general form of radiative transfer equation in spherically symmetric atmospheres relevant to relativistic flow. $0 \leq \beta \leq 1$.

For most motions of astrophysical interest $(V/c) \ll 1$, and it is adequate to retain terms upto $O(V/c)$ in the radiative transfer equations. Under these circumstances, the transfer equation (2.5.18) reduces to

$$\frac{1}{c}\left[\frac{\partial}{\partial t}+V\frac{\partial}{\partial r}\right]I^0(\mu_0,\nu_0) + \frac{\mu_0}{r^2}\frac{\partial}{\partial r}[r^2 I^0(\mu_0,\nu_0)]$$
$$+ \frac{\partial}{\partial \mu_0}\left[(1-\mu_0^2)\left\{\frac{1}{r}+\frac{\mu_0}{c}(\frac{V}{r}-\frac{\partial V}{\partial r})-\frac{a}{c^2}\right\}I^0(\mu_0,\nu_0)\right]$$
$$- \frac{\partial}{\partial \nu_0}\left[\nu_0\left\{(1-\mu_0^2)\frac{V}{cr}+\frac{\mu_0^2}{c}\frac{\partial V}{\partial r}+\frac{\mu_0 a}{c^2}\right\}I^0(\mu_0,\nu_0)\right]$$
$$+ \left[(3-\mu_0^2)\frac{V}{cr}+\frac{1+\mu_0^2}{c}\frac{\partial V}{\partial r}+\frac{2\mu_0 a}{c^2}\right]I^0(\mu_0,\nu_0)$$
$$= \eta^0(\nu_0) - \chi^0(\nu_0)I^0(\mu_0,\nu_0), \qquad (2.5.19)$$

where $a = \frac{\partial V}{\partial t}$; the acceleration of the fluid.

2.6 Moment Equations in Comoving Frame

Multiplying (2.5.18) by $(d\Omega_0/4\pi)$ and $(\mu_0 d\Omega_0/4\pi)$ and integrating over angles, the angular moment equations give us the monochromatic energy and momentum equations respectively. The monochromatic radiation energy equation is given by

$$\gamma\left[\frac{\partial E^0(\nu_0)}{\partial t}+\frac{V}{c^2}\frac{\partial \tilde{F}^0(\nu_0)}{\partial t}\right]+\gamma\left[\frac{\partial \tilde{F}^0(\nu_0)}{\partial r}+V\frac{\partial E^0(\nu_0)}{\partial r}\right]$$
$$+ \gamma\left[\frac{1}{r}\{2\tilde{F}^0(\nu_0)+3VE^0(\nu_0)-VP^0(\nu_0)\}\right.$$
$$+ \gamma^2\frac{\partial V}{\partial r}\frac{2}{c^2}V\tilde{F}^0(\nu_0)+E^0(\nu_0)+P^0(\nu_0)$$
$$\left.+ \frac{\gamma^2}{c^2}\frac{\partial V}{\partial t}\{2\tilde{F}^0(\nu_0)+VE^0(\nu_0)+VP^0(\nu_0)\}\right]$$

2. Radiation Field and Transfer Equations in Moving Media

$$
\begin{aligned}
& -\frac{\partial}{\partial \nu_0}\left[\gamma \nu_0\left\{\frac{V}{r}(E^0(\nu_0) - P^0(\nu_0))\right.\right. \\
& \left.+ \gamma^2 \frac{\partial V}{\partial r}(P^0(\nu_0) + \frac{V}{c^2}\tilde{F}^0(\nu_0))\right. \\
& \left.\left.+ \frac{\gamma^2}{c^2}\frac{\partial V}{\partial t}(VP^0(\nu_0) + \tilde{F}^0(\nu))\right\}\right] \\
& = 4\pi\eta^0(\nu_0) - c\chi^0(\nu_0)E^0(\nu_0),
\end{aligned}
\qquad (2.6.1)
$$

where

$$E^0(\nu_0) = \frac{1}{c}\oint I^0(\mathbf{r},t;\tilde{\mathbf{s}}_0,\nu_0)d\Omega_0 = \frac{2\pi}{c}\int_{-1}^{1} I^0 d\mu_0. \qquad (2.6.2a)$$

$$\tilde{F}^0(\nu_0) = \oint I^0(\mathbf{r},t;\tilde{\mathbf{s}}_0,\nu_0)\tilde{\mathbf{s}}_0 d\Omega_0 = 2\pi\int_{-1}^{1} I^0 \mu_0 d\mu_0. \qquad (2.6.2b)$$

$$P^0(\nu_0) = \frac{1}{c}\oint I^0(\mathbf{r},t;\tilde{\mathbf{s}}_0,\nu_0)\tilde{\mathbf{s}}_0\tilde{\mathbf{s}}_0 d\Omega_0 = \frac{2\pi}{c}\int_{-1}^{1} I^0 \mu_0^2 d\mu_0. \qquad (2.6.2c)$$

Similarly, the monochromatic momentum equation can be written down as

$$
\begin{aligned}
& \frac{\gamma}{c^2}\left[V\frac{\partial P^0(\nu_0)}{\partial t} + \frac{\partial \tilde{F}^0(\nu_0)}{\partial t}\right] + \gamma\left[\frac{V}{c^2}\frac{\partial \tilde{F}^0(\nu_0)}{\partial r} + \frac{\partial P^0(\nu_0)}{\partial r}\right] \\
& + \gamma\left[\frac{1}{r}\left\{\frac{2V}{c^2}\tilde{F}^0(\nu_0) - E^0(\nu_0) + 3P^0(\nu_0)\right\}\right. \\
& + \frac{\gamma^2}{c^2}\frac{\partial V}{\partial r}\{2\tilde{F}^0(\nu_0) + VE^0(\nu_0) + VP^0(\nu_0)\} \\
& \left.+ \frac{\gamma^2}{c^2}\frac{\partial V}{\partial t}\left\{\frac{2V}{c^2}\tilde{F}^0(\nu_0) + E^0(\nu_0) + P^0(\nu_0)\right\}\right] \\
& - \frac{\partial}{\partial \nu_0}\left[\gamma \nu_0\left\{\frac{V}{c^2 r}(\tilde{F}^0(\nu_0) - L^0(\nu_0))\right.\right. \\
& + \frac{\gamma^2}{c^2}\frac{\partial V}{\partial r}(L^0(\nu_0) + VP^0(\nu_0)) \\
& \left.\left.+ \frac{\gamma^2}{c^2}\frac{\partial V}{\partial t}(P^0(\nu_0) + \frac{V}{c^2}L^0(\nu_0))\right\}\right] \\
& = -(\frac{\chi^0(\nu_0)}{c})\tilde{F}^0(\nu_0),
\end{aligned}
\qquad (2.6.3)
$$

where
$$L^0(\nu_0) = 2\pi \int_{-1}^{1} I^0(\mu_0, \nu_0)\mu_0{}^3 d\mu_0. \qquad (2.6.4)$$

The term L^0 did not occur in the lab frame monochromatic momentum equation.

The total energy equation is obtained by integrating (2.6.1) over comoving frame frequencies and is given by

$$\gamma\left[\frac{\partial E^0}{\partial t} + \frac{V}{c^2}\frac{\partial \tilde{F}^0}{\partial t}\right] + \gamma\left[\frac{\partial \tilde{F}^0}{\partial r} + V\frac{\partial E^0}{\partial r}\right]$$
$$+ \gamma\left[\frac{1}{r}\{2\tilde{F}^0 + 3VE^0 - VP^0\}\right.$$
$$+ \gamma^2\frac{\partial V}{\partial r}\left\{\frac{2}{c^2}V\tilde{F}^0 + E^0 + P^0\right\}$$
$$\left.+ \frac{\gamma^2}{c^2}\frac{\partial V}{\partial t}\{2\tilde{F}^0 + VE^0 + VP^0\}\right]$$
$$= \int_0^\infty [4\pi\eta^0(\nu_0) - c\chi^0(\nu_0)E^0(\nu_0)]d\nu_0. \qquad (2.6.5)$$

Similarly the total momentum equation is obtained by integrating (2.6.3) over all frequencies in comoving frame and is given by

$$\frac{\gamma}{c^2}\left[V\frac{\partial P^0}{\partial t} + \frac{\partial \tilde{F}^0}{\partial t}\right] + \gamma\left[\frac{V}{c^2}\frac{\partial \tilde{F}^0}{\partial r} + \frac{\partial P^0}{\partial r}\right]$$
$$+ \gamma\left[\frac{1}{r}\left\{\frac{2V}{c^2}\tilde{F}^0 - E^0 + 3P^0\right\}\right.$$
$$+ \frac{\gamma^2}{c^2}\frac{\partial V}{\partial r}\{2\tilde{F}^0 + VE^0 + VP^0\}$$
$$\left.+ \frac{\gamma^2}{c^2}\frac{\partial V}{\partial t}\left\{\frac{2V}{c^2}\tilde{F}^0 + E^0 + P^0\right\}\right]$$
$$= -\frac{1}{c}\int_0^\infty \chi^0(\nu_0)\tilde{F}^0(\nu_0)d\nu_0. \qquad (2.6.6)$$

It is to be noted that L^0 has dropped out of the equation.

The equations (2.6.1) to (2.6.6) are applicable to radiative transfer problems in media with high velocity fields. However in most flows of astrophysical interest $(V/c) \ll 1$ and it is sufficient to retain terms only upto $O(V/c)$ For this we follow a procedure similar to that for obtaining (2.6.1) and (2.6.2) from (2.5.18). We have for the monochromatic radiative energy equation, retaining terms upto $O(V/c)$,

$$\frac{DE(\nu_0)}{Dt} + \frac{1}{r^2}\frac{\partial}{\partial r}[r^2 \tilde{F}^0(\nu_0)] + \frac{V}{r}[3E^0(\nu_0) - P^0(\nu_0)]$$
$$+ \frac{\partial V}{\partial r}[E^0(\nu_0) + P^0(\nu_0)] + \frac{2a}{c^2}\tilde{F}^0(\nu_0)$$
$$- \frac{\partial}{\partial \nu_0}\left[\nu_0\left\{\frac{V}{r}(E^0(\nu_0) - P^0(\nu_0)) + \frac{\partial V}{\partial r}P^0(\nu_0) + \frac{a}{c^2}\tilde{F}^0(\nu_0)\right\}\right]$$
$$= 4\pi\eta^0(\nu_0) - c\chi^0(\nu_0)E^0(\nu_0) \qquad (2.6.7)$$

and for monochromatic radiative momentum equation

$$\frac{1}{c^2}\frac{D\tilde{F}^0(\nu_0)}{Dt} + \frac{\partial P^0(\nu_0)}{\partial r} + \frac{1}{r}[3P^0(\nu_0) - E^0(\nu_0)]$$
$$+ \frac{2}{c^2}\left(\frac{\partial V}{\partial r} + \frac{V}{r}\right)\tilde{F}^0(\nu_0) + \frac{a}{c^2}[E^0(\nu_0) + P^0(\nu_0)]$$
$$- \frac{\partial}{\partial \nu_0}\left[\nu_0\left\{\frac{V}{c^2 r}(\tilde{F}^0(\nu_0) - L^0(\nu_0)) + \frac{1}{c^2}\frac{\partial V}{\partial r}L^0(\nu_0)\right.\right.$$
$$\left.\left. + \frac{a}{c^2}P^0(\nu_0)\right\}\right]$$
$$= -\frac{1}{c}\chi^0(\nu_0)\tilde{F}^0(\nu_0). \qquad (2.6.8)$$

In (2.6.7) and (2.6.8) (D/Dt) stands for comoving frame time derivative often known as Lagrangian or substantive derivative. Explicitly

$$\frac{DA}{Dt} = \frac{\partial A}{\partial t} + (\mathbf{V}\cdot\nabla)A \qquad (2.6.9)$$

where A can be a scalar, a vector or a tensor.

Finally integrating over all frequencies in the comoving frame we get the total radiation energy and total momentum equations as

$$\frac{DE^0}{Dt} + \frac{1}{r^2}\frac{\partial}{\partial r}(r^2 \tilde{F}^0) + \frac{V}{r}(3E^0 - P^0)$$
$$+ \frac{\partial V}{\partial r}[E^0 + P^0] + \frac{2a}{c^2}\tilde{F}^0$$
$$= \int_0^\infty [4\pi \eta^0(\nu_0) - c\chi^0(\nu_0)E^0(\nu_0)]d\nu_0, \qquad (2.6.10)$$

and

$$\frac{1}{c^2}\frac{D\tilde{F}^0}{Dt} + \frac{\partial P^0}{\partial r} + \frac{1}{r}[3P^0 - E^0]$$
$$+ \frac{2}{c^2}(\frac{\partial V}{\partial r} + \frac{V}{r})\tilde{F}^0 + \frac{a}{c^2}[E^0 + P^0]$$
$$= -\frac{1}{c}\int_0^\infty \chi^0(\nu_0)\tilde{F}^0(\nu_0)d\nu_0. \qquad (2.6.11)$$

The equations (2.5.19), (2.6.6) -(2.6.8), (2.6.10) and (2.6.11) are equivalent to those obtained by Castor (1972), Buchler (1979). Castor, however, omitted the acceleration term (which is of the order of $O(V/c)^2$ in the fluid flow time scale). Buchler also gave the results for radiative transfer in a medium with cylindrical symmetry.

References

Buchler, J.R. (1979): JQSRT, **22**, 293.
Castor, J. (1972): Ap. J., **178**, 779.
Mihalas, D. (1978): Stellar Atmospheres. W. Freeman and Co., San Francisco.
Mihalas, D., Mihalas, B. W. (1984): Foundations of Radiation Hydrodynamics. Oxford University Press, New York.
Ozisik, M. N. (1973): Radiative Transfer. Wiley Interscience Publications, New York.
Sen, K.K., Wilson, S J. (1990): Radiative Transfer in Curved Media. World Scientific, Singapore.

PART

II

Chapter 3. Transfer Equations in Lab Frame and Their Solutions

In the early half of twentieth century, it was realised that certain class of stars such as WR-, Be, P-cygni etc exhibit abnormal structure of spectral lines. One common feature of them was the presence of emission lines in their spectra. A search for physical models of such stars which describe the states of their constituent material and reflect the methods of formation of these lines was immediately on. The following basic properties were agreed upon for the model.

(a) All these stars have extended atmospheres.

(b) The lower layers of the envelope have the properties of ordinary stellar atmosphere and in it originate the absorption lines and continuum spectra.

(c) The outer layers of the atmosphere are of low density and have relatively cooler temperature. The emission lines originate in these outer layers.

(d) The outermost layers of atmosphere are transparent except in the frequencies of the emission lines.

(e) The abnormal broadening of spectral lines in these stars suggests that their atmospheres are in some sort of motion such as expansion and/or rotation.

In the present chapter, we shall state the early attempts at explaining the broad features of the spectra of stars with moving atmospheres. We shall mainly confine ourselves to the study of the role of radiative transfer in the line formation of these stars. Most of the theoretical work done at the early stages were of the diagnostic character designed to explain the special spectral structure of each of the class of stars WR-, Be or P-cygni.

78   3. Transfer Equations and Solutions in Lab Frame

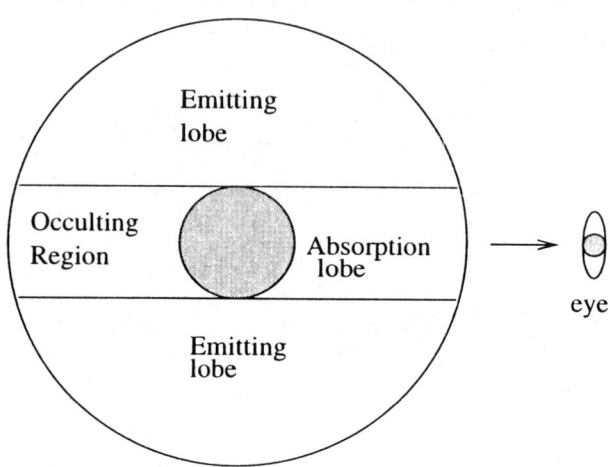

Figure 3.1. Schematic structure of a Be-shell star rotating envelope — a view from the pole

The following physical processes were suggested to explain the characteristic shapes of emission lines exhibited by them.

For WR-stars, Beals (1929–1950) took note of the similarity of WR-spectra and that of novae. He proposed accretion or ejection of matter as the mechanism responsible for generating the emission lines. Münch (1950) considered the broadening due to electron scattering. Thomas (1949) attributed it to the non-thermal turbulent motion of the atmospheric material.

Miyamoto (1941, 1952) and Kogure (1959–69) studied in detail the excitation and ionisation equilibria in the atmospheres of Be-stars. They concluded that the conditions in the WR-atmospheres resemble those in the Be-shell stars. Miyamoto's method was a static method.

Sobolev (1957, 1960) proposed a comprehensive theory for solving radiative transfer problems in the envelopes moving around WR-, Be- and P-cygni stars. His method is known as the "escape probability method". It is based on the inquiry into the physical effects of the nature of motion occurring in the atmospheres of stars such as expansion, rotation or accretion of matter. Direct reference to the radiative transfer equation is avoided.

In what follows, we shall describe the essential characteristics of the "escape probability method" of Sobolev, the static method of Miyamoto and Kogure and also refer to the method using free electron diffusion as the mechanism. Of these, Sobolev's escape probability method is still holding ground. The other two methods are of historical importance in the evolution of methodologies for solving radiative transfer problems in spherically symmetric moving media.

In the following figures, we show the qualitative features of the spectral lines of WR-, P-cygni and Be stars.

(a) WR-star.

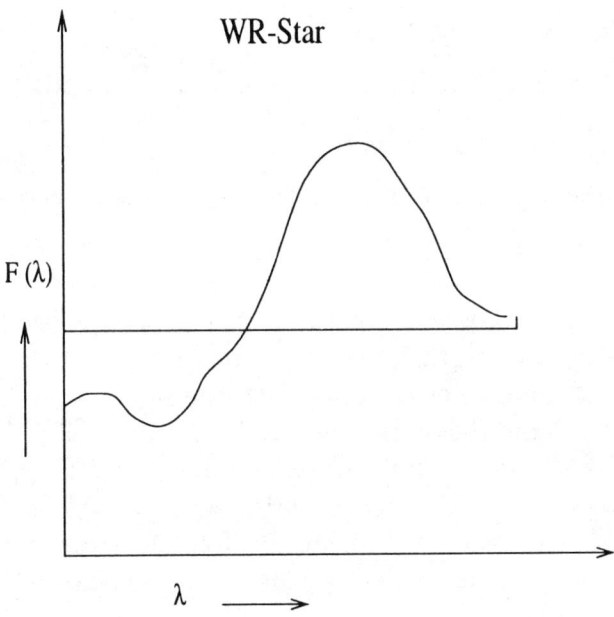

Figure 3.2. Schematic diagram of the profile of a representative WR-star

(b) P-cygni star.

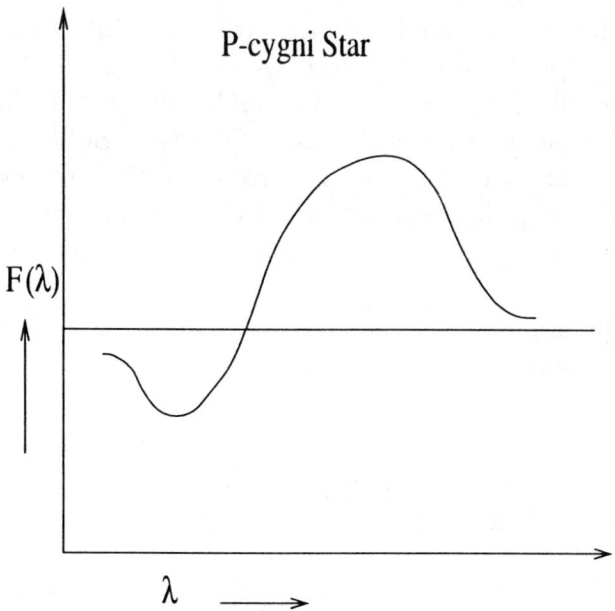

Figure 3.3. Schematic diagram of H_α profile of a representative P-cygni star

The study of the spectral features of each of these stars involves the understanding of star's envelope dynamics and the radiation field. We shall be concerned only with the latter.

We shall outline below the distinctive features of Sobolev's escape probability method which uses differential Doppler effect, the static method of Miyamoto and Kogure based on Rosseland Cycle and the method of diffusion by free electrons proposed by Schuster (1905), Underhill (1949), Burbridge (1953–54) and others.

(c) Be-star.

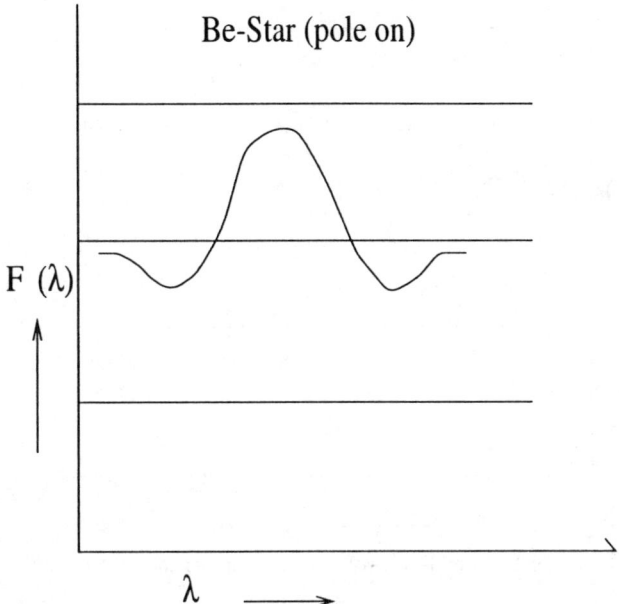

Figure 3.4. Schematic Balmer line profile of a representative Be (pole on) star

The Differential Doppler Effect: Escape Probability Method of Sobolev

In a moving medium, in the presence of a velocity gradient, photons emitted at one point of the medium are Doppler shifted when they arrive at another position. The amount of this Doppler shift is proportional to the product of average velocity gradient and the distance between the two positions. As a consequence, the photons emitted at some position of the medium at say the line centre, can interact only with the material confined to a region localised around its origin. Beyond this they fall too far into the wing of the line profile of the material at the other position. Hence they escape interaction with the material lying beyond and they cannot be absorbed. Note should also be taken of the fact that difference of frequency between two positions depends not only on

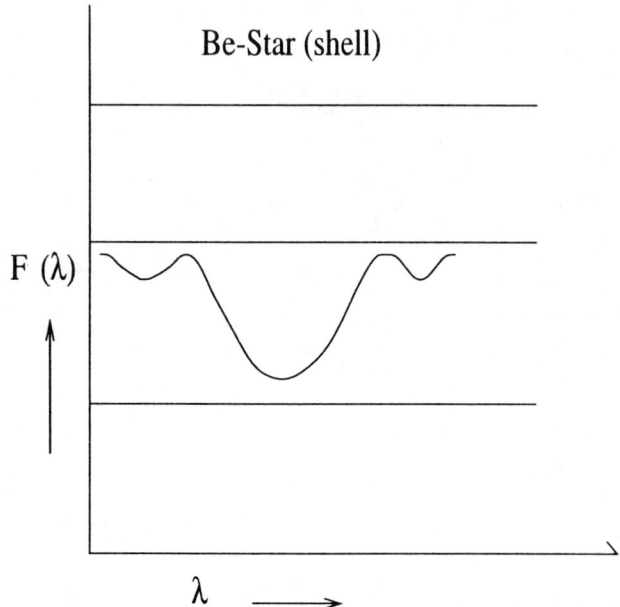

Figure 3.5. Schematic Balmer line profile of a representative Be-shell star

the magnitude of the relative velocity between the two positions but also on the direction.

Sobolev's theory takes account of the escape probability of the photons emitted from different positions of the medium. In the following, we shall illustrate Sobolev's theory with reference to the radiative transfer in a spherically symmetric radially expanding model of a stellar atmosphere.

The qualitative model of a star enveloped by a spherically symmetric expanding atmosphere is as follows. A central star is surrounded by a radially expanding atmosphere, the star's photosphere being well defined. It is assumed that the atmosphere is transparent in the sense that every photon emitted in it is received by the observer. The line emission and absorption is supposed to take place at the line centre frequency. An external observer receives the line centre frequency Doppler shifted by the velocity of the material of the atmosphere along the line of sight. The velocity along the line of sight may be positive or negative.

3. Transfer Equations and Solutions in Lab Frame   83

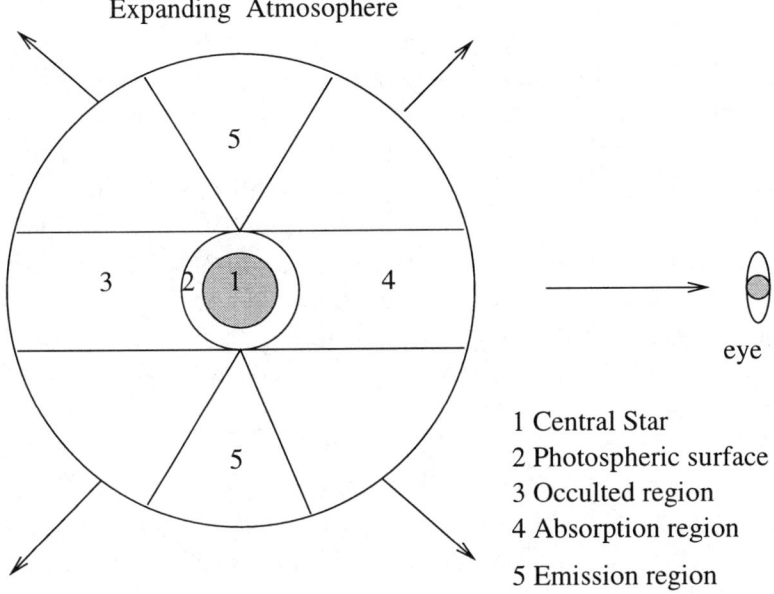

Figure 3.6. Schematic diagram of an expanding envelope surrounding a central star

Sobolev avoided formal solution of radiative transfer equation and based his theory principally on the physical study of the transport of photons through the expanding envelopes of the stars [cf. Fig. 3.6].

In Sobolev's theory, the idea is to study the process of line formation in an optically thick expanding spherical envelope surrounding an opaque core star of radius r_c. He ignores the existence of background continuum. A fluid element in the envelope is selected. An observer is placed in a coordinate system at rest with respect to the test fluid element. The observer looks along the ray and observes a differential Doppler shift of each successive sample point along the ray with respect to the chosen test point. The Doppler shift is due to the existence of a velocity gradient in the envelope. Eventually its effect may be so large that the interaction between the photons in the line profile at the test point and the other sample points is not possible. The existence of a velocity field thus offers a sort of escape mechanism for the photons beyond the interaction limit. They are no more absorbed and they escape freely to infinity. When the velocity gradient is large, the interaction region becomes small and in this region the physical properties like temperature, density, ionisation state etc can be taken to be homogeneous. And on this realisation, the theories can be proposed on the basis of local quantities and a photon escape probability parameter $\mathcal{P}_e$. $\mathcal{P}_e$ is the measure of the probability of photon escape summed over all directions and line frequencies.

An excellent review of Sobolev's theory is given in Mihalas (1978, p. 471–490). We shall outline below the main features of the method. Rybicki (1984) discussed the different scopes of the escape probability method, its possible modifications and applications to static and moving media, in particular those with high differential velocity. He noted that Sobolev's theory gives the escape probability parameter as a function of the local conditions of the gas. Following the discussion of Sobolev's method we shall put forward the comments of Rybicki.

3.1 The Escape Probability Method of Sobolev

The conventional definition of escape probability is the measure of the probability of a photon emitted at a test point escaping the medium at a single flight without suffering absorption or scattering in its way. We consider a photon in the medium at a particular point with its frequency and direction specified. Then the escape probability p_ν, of a photon of frequency ν is given by the exponential extension law as

$$p_\nu = \exp(-[\tau_\nu]_f), \qquad (3.1.1)$$

where $[\tau_\nu]_f$ is the optical depth along the ray in the forward direction for the frequency ν. When averaged over frequencies emitted along the ray and the angles, we get the photon escape probability parameter $\mathcal{P}_e$ as

$$\mathcal{P}_e = \frac{1}{4\pi} \oint d\Omega \int_0^\infty \phi(\nu)\, p_\nu\, d\nu, \qquad (3.1.2)$$

where $\phi(\nu)$ is the line profile.

For normal lines the optical depth is positive and $0 \leq \mathcal{P}_e \leq 1$.

In Sobolev's theory, the velocity gradient in the medium is assumed large. The largeness of the velocity gradient is expressed in terms of a quantity L, called the Sobolev length. L is defined as the length over which the profile function of a line is shifted through a distance equal to its own width by the macroscopic velocity gradient existing in the medium. We denote the line profile function at frequency ν as $\phi_{ij}(\nu)$ for the line $i \to j$ and this is normalised as

$$\int_0^\infty \phi_{ij}(\nu)\, d\nu = 1. \qquad (3.1.3)$$

Let ν_{ij} be the line centre frequency in the lab frame and let the frequency be expressed in the dimensionless form as

$$x = \frac{\nu - \nu_{ij}}{\Delta\nu_D} \qquad (3.1.4)$$

where $\Delta\nu_D$ is the Doppler width given by

$$\Delta\nu_D = \frac{v_{th}}{c}\, \nu_{ij}. \qquad (3.1.5)$$

v_{th} is the thermal velocity.

We recall that (cf. 2.1.1) with $\nu' = \nu_0$

$$\nu_{ij} = \nu_0 \left[1 + \frac{1}{c}\, \mathbf{n} \cdot \mathbf{V}(r)\right] = \nu_0 \left(1 + \frac{v_s}{c}\right), \qquad (3.1.6)$$

where ν_0 is the line centre frequency in the comoving frame and v_s is the component of the velocity field along the direction of the ray, i.e. in this case the line of sight.

Then

$$\delta\nu_{ij} = (\nu_0/c) \left(\frac{dv_s}{ds}\right) \delta s. \qquad (3.1.7)$$

The Sobolev length L is defined to be δs for which $\delta\nu_{ij} = \Delta\nu_D$.
Hence from (3.1.7) and (3.1.5)

$$L = \frac{c\Delta\nu_D}{\nu_0 \left|\frac{dv_s}{ds}\right|} = \frac{v_{th}\nu_0(1+\frac{v_s}{c})}{\nu_0 \left|\frac{dv}{ds}\right|} \approx v_{th} \bigg/ \left|\frac{dv_s}{ds}\right|. \qquad (3.1.8)$$

Note that L is the same for all lines.

Let R be the distance by which the other macroscopic quantities of the radiation field varies. Sobolev's theory is applicable when

$$L \ll R. \tag{3.1.9}$$

If we estimate $\frac{dv_s}{ds}$ to be V/R where V is the velocity of the medium then

$$v_{th} \ll V. \tag{3.1.10}$$

Since v_{th} is of the order of velocity of sound, the Sobolev's method is sometimes termed as supersonic approximation.

A photon emitted at a test point travels through L to a point A, say. The profile at that point has Doppler shifted by $\Delta\nu_D$. The photon can then escape the local neighbourhood of the test point without any interaction and obstruction. Furthermore, if the velocity gradient does not change sign along the ray, the photon will escape the atmosphere itself.

Taking the above factors into account Sobolev enunciated a scheme for obtaining the radiation field which we outline below.

Let $\bar{J}(r)$ be the mean intensity averaged over a line profile defined by

$$\bar{J}(r) = \int J_\nu \, \phi(\nu) d\nu. \tag{3.1.11}$$

Then from physical reasoning, we may write

$$\bar{J}(r) = (1 - \mathcal{P}_e)S(r) + \mathcal{P}_c I_c, \tag{3.1.12}$$

where $\mathcal{P}_e$ is the photon escape probability,

$S(r)$, the frequency integrated source function
$\mathcal{P}_c$, the probability of penetration of photon integrated over angle and frequency of the specific intensity I_c emitted from the source to the test point.

When $\mathcal{P}_e = 0$ and $\mathcal{P}_c = 0$, we obviously have

$$\bar{J}(r) = S(r). \tag{3.1.13}$$

From (3.1.12) it is clear that we can find $\bar{J}(r)$ when $\mathcal{P}_e$ and $\mathcal{P}_c$ are known.

Determination of $\mathcal{P}_e$ and $\mathcal{P}_c$

We shall try to define $\mathcal{P}_e$ and $\mathcal{P}_c$ in terms of local opacity and velocity gradient. We express velocity and frequency in dimensionless units.

Let
$$V(r) = (v(r)/v_{th}) \tag{3.1.14}$$
and
$$x = \frac{\nu - \nu_0}{\triangle \nu_D} \tag{3.1.15}$$

where v_{th} is the thermal velocity and $\triangle \nu_D$, the Doppler width given by $\triangle \nu_D = (v_{th}/c)\nu_0$.

In spherical coordinates, $I(r,\theta,\nu)$ is replaced by $I(s,p,x)$ and so also the other quantities representing the radiation field. The relation between $P(s,p,x)$ and $P(r,\theta,\nu)$ is shown in Fig. 3.7.

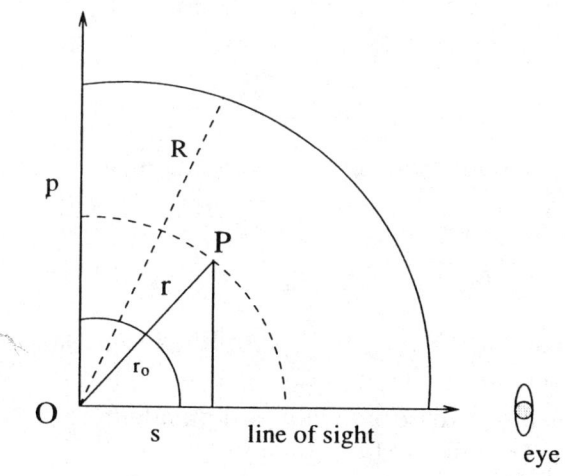

Figure 3.7. The p–s coordinate system for solving the transfer equation in spherical geometry

3. Transfer Equations and Solutions in Lab Frame

We have
$$\left.\begin{array}{l} s = r\cos\theta = r\mu \\ p = r\sin\theta = r(1-\mu^2)^{1/2}. \end{array}\right\} \quad (3.1.16)$$

The positive direction of s is the direction of line of sight towards the observer. Then the optical thickness along the s-direction is given by
$$\begin{aligned} \tau(s,p,x) &= \int_s^\infty \chi(s',p,x)\,ds' \\ &= \int_s^\infty \chi_\ell(r')\phi(x')\,ds'. \end{aligned} \quad (3.1.17)$$

Here
$$\left.\begin{array}{l} r' = (s'^2 + p'^2)^{1/2}, \quad \mu' = (s'/r') \\ x' = x - V_s(s') = x - \mu V(r) \end{array}\right\} \quad (3.1.18)$$

The profile function $\phi(x')$ contributes most near $x' = 0$. From (3.1.18) when $x' = 0$, we have
$$V_s = V(r)\cos\theta = V(r)\mu = x.$$

Thus the surface of constant radial velocity ($V_{s_0} = x$) define the point s_0. If r_0, is the corresponding radial coordinate for a given p then
$$r_0 = (s_0^2 + p^2)^{1/2}.$$

Since the main contribution comes from a surface of constant radial velocity we can approximate $\chi_\ell(r')$ by $\chi_\ell(r_0)$ in (3.1.17) and write
$$\tau(s,p,x) = \chi_\ell(r_0)\int_s^\infty \phi(x')ds'.$$

In some cases (3.1.17) can be simplified without affecting the results adversely by replacing $\chi_\ell(r')$ by $\chi_\ell(r_0)$ and taking it outside the integrals. This obviously is equivalent to assuming constancy of opacity within the range of s.

Now in (3.1.17), we change the variable from s' to x' through (3.1.18).

From (3.1.18), the velocity gradient $(\partial V_s/\partial s)_p$ is given by
$$(\partial V_s/\partial s)_p = -(\partial x'/\partial s)_p = [\partial\{\mu V(r)\}/\partial s]_p. \quad (3.1.19)$$

We remember that $\mu = \mu(s,p)$ and $r = r(s,p)$ and note that $\frac{\partial r}{\partial s} = \frac{s}{r} = \mu$ and $\frac{\partial \mu}{\partial s} = \frac{1-\mu^2}{r}$. Hence (3.1.19) can be written as

3.1 The Escape Probability Method of Sobolev

$$(\partial V_s/\partial s)_p = \left\{\mu \frac{\partial V(r)}{\partial s} + V(r)\frac{\partial \mu}{\partial s}\right\}$$

$$= \left\{\mu^2 \frac{\partial V}{\partial r} + (1-\mu^2)\frac{V}{r}\right\}$$

$$= \left(\frac{V}{r}\right)\left[(1-\mu^2) + \mu^2\left(\frac{r}{V}\right)\frac{\partial V}{\partial r}\right]$$

$$= Q(r,\mu)\text{say}. \qquad (3.1.20)$$

We define $\Phi(x)$ as

$$\Phi(x) = \int_{-\infty}^{x} \phi(\xi')\,d\xi', \qquad (3.1.21)$$

and we write equation (3.1.17)

$$\tau(s,p,x') = \tau(-\infty,p,x)\,\Phi(x') \qquad (3.1.22)$$

where $x' = x'(s,p,x)$ is given by

$$x' = x - V_s(s') = x - \mu V(r).$$

Then from (3.1.17), (3.1.20) and (3.1.21)

$$\tau(-\infty,p,x) = \int_{-\infty}^{\infty} \chi_\ell(r')\,\phi(x')\,ds'$$

$$= \chi_\ell(r_0) \int_{-\infty}^{\infty} \phi(x')\frac{\partial s'}{\partial x'}\,dx'$$

$$= \chi_\ell(r_0) \int_{-\infty}^{\infty} \phi(x')\,dx'/Q(r_0,\mu_0)$$

$$= \chi_\ell(r_0)/Q(r_0,\mu_0). \qquad (3.1.23)$$

From (3.1.21), we have

$$\Phi(-\infty) = 0 \quad \Phi(\infty) = 1. \qquad (3.1.24)$$

Hence (3.1.23) can be written as

$$\tau(-\infty,p,x) = [\chi_\ell(r_0)/(V/r_0)]\bigg/\left\{(1-\mu^2) + \mu^2\left(\frac{r_0}{V}\right)\frac{\partial V}{\partial r}\right\}$$

$$= \tau_0(r_0)\bigg/\left[1 + \mu^2\left\{\frac{d\ell n V}{d\ell n r_0} - 1\right\}\right] \qquad (3.1.25)$$

where
$$\tau_0(r_0) = \chi_\ell(r_0)/(V/r_0). \qquad (3.1.26)$$

In writing equations (3.1.17) to (3.1.26), we have assumed that the region of interaction is small and the coefficients of transformation are evaluated at $s_0(p, x)$ or $r_0(s_0, p_0)$, $\mu_0(s_0, p_0)$.

$\chi_\ell(r_0)$ can be written in terms of atomic variables as

$$\chi_\ell(r_0) = c_{ij}\left[n_i(r_0) - \left(\frac{g_i}{g_j}\right)n_j(r_0)/\triangle \nu_D\right] \qquad (3.1.27)$$

where
$$c_{ij} = (\pi e^2 mc) f_{ij}.$$

f_{ij} is the oscillator strength of $i \to j$ transition and c_{ij}, the total scattering cross-section. g_i, g_j are the statistical weights of atomic states i and j. e is the electronic charge and m, the mass of electron.

To Find $\mathcal{P}_e(r)$

We recall that $\mathcal{P}_e(r)$ is the measure of the escape probability of photon from the point r in the medium summed over all directions and frequencies. From (3.1.1) and (3.1.2), assuming the spherical symmetry of the radiation field, we have

$$\mathcal{P}_e(r) = \frac{1}{2}\int_{-1}^{1} d\mu \int_{-\infty}^{\infty} \phi(x')\,\exp\left\{-\tau(s, p, x')\right\} dx', \qquad (3.1.28)$$

where $s = s(r, \mu)$, $p = p(r, \mu)$, $x' = x'(s, p, x)$.

When the material in the region of interaction is homogeneous, the distinction between r and r_0 may be avoided. In the following we shall assume it to be so. Furthermore, we shall assume the core star to be opaque to the impinging photons.

Then using (3.1.20) to (3.1.28), we have

$$\begin{aligned}\mathcal{P}_e(r) &= \frac{1}{2}\int_{-1}^{1} d\mu \int_{0}^{1} \exp\left\{-\chi_\ell(r)\,\Phi/Q(r,\mu)\right\} d\Phi \\ &= [\chi_\ell(r)]^{-1} \int_{0}^{1} [1 - \exp\left\{-\chi_\ell(r)/Q(r,\mu)\right\} Q(r,\mu)]\, d\mu.\end{aligned}$$
$$(3.1.29)$$

3.1 The Escape Probability Method of Sobolev

In the special case of the velocity being directly proportional to r, i.e. $V = kr$, k a constant we have

$$Q(r,\mu) = \mu^2 \frac{\partial V}{\partial r} + (1-\mu^2)\frac{V}{r} = k. \quad (3.1.30)$$

Then from (3.1.29)

$$\begin{aligned}\mathcal{P}_e(r) &= \frac{k}{\chi_\ell(r)} \int_0^1 [1 - \exp\{-\chi_\ell(r)/k\}]\, d\mu \\ &= [1 - \exp\{-\tau_0(r)\}]/\tau_0(r),\end{aligned} \quad (3.1.31)$$

where

$$\tau_0(r) = \chi_\ell(r)/k. \quad (3.1.32)$$

Thus we see that when the velocity is directly proportional to r, $\mathcal{P}_e(r)$ takes a simple form. As long as $Q(r,\mu)$ is constant (3.1.31) will hold good.

To Find $\mathcal{P}_c$

We recall that $\mathcal{P}_c$ is defined to be the probability of penetration of photon of specific intensity I_c to the test point r, the photons being emitted from the core, and this probability being integrated over frequencies and angles. We take the test point r at a large distance from the core surface. The outer surface of the atmosphere is assigned zero at ∞.

Then $\mathcal{P}_c$ can be written as

$$\begin{aligned}\mathcal{P}_c &= \frac{1}{2}\int_{-1}^{-\mu_c} d\mu \int_0^1 d\Phi\, \exp[-\chi_\ell(r)\,\Phi/Q(r,\mu)] \\ &= \frac{1}{2\chi_\ell(r)} \int_{\mu_c}^1 [1 - \exp\{-\chi_\ell(r)/Q(r,\mu)\}]\, Q(r,\mu) d\mu\end{aligned} \quad (3.1.33)$$

where

$$\mu_c = \left(1 - \frac{r_c^2}{r^2}\right)^{1/2}. \quad (3.1.34)$$

When the velocity field in the medium is given by $V = kr$ with, k constant, [cf. 3.1.30]

$$\begin{aligned}
\mathcal{P}_c &= \frac{k}{2\chi_\ell(r)} \int_{\mu_c}^{1} [1 - \exp\{-\chi_\ell(r)/k\}] \, d\mu \\
&= \frac{k}{2\chi_\ell(r)} [\{1 - \exp(-\chi_\ell(r)/k)\} - \mu_c\{1 - \exp(-\chi_\ell(r)/k)\}] \\
&= \frac{1}{2} \mathcal{P}_e(r)(1 - \mu_c) \\
&= \frac{1}{2} \mathcal{P}_e \left[1 - (1 - \frac{r_c^2}{r^2})^{1/2} \right] \\
&= W \mathcal{P}_e(r), \quad\quad\quad\quad\quad\quad\quad (3.1.35)
\end{aligned}$$

where W, the dilution factor is given by

$$W = \frac{1}{2} \left[1 - (1 - \frac{r_c^2}{r^2})^{1/2} \right]. \quad\quad (3.1.36)$$

We note that $\mathcal{P}_e$ and $\mathcal{P}_c$ are dependent on the local values of velocity gradient and opacity at r. When $\mathcal{P}_e$ and $\mathcal{P}_c$ are known, we can calculate the mean intensity $\bar{J}(r)$ from (3.1.12).

Sobolev's method of escape probability of photons is simple and elegant. It succeeds in obtaining the mean intensity without going through the process of solving the equation of radiative transfer.

The expression for the source function $S(r)$ for two-level atoms is given by

$$S(r) = \left[(1 - \varepsilon)\bar{J} + \varepsilon B \right] \quad\quad (3.1.37)$$

where $\varepsilon = \varepsilon'/(1 + \varepsilon')$ [cf. 1.4.17].

In writing (3.1.37) complete redistribution of frequency is assumed. B is the Planck function and ε is the fraction of line emission which is of thermal origin.

From (3.1.12)

$$\bar{J}(r) = (1 - \mathcal{P}_e)S(r) + \mathcal{P}_c I_c.$$

Eliminating $\bar{J}$ from (3.1.37) and (3.1.12), we have

$$S(r) = [(1 - \varepsilon)\mathcal{P}_e I_c + \varepsilon B] / [(1 - \varepsilon)\mathcal{P}_e + \varepsilon]. \quad\quad (3.1.38)$$

Thus when $\mathcal{P}_e$ and $\mathcal{P}_c$ are known, we can determine $S(r)$.

Flux and Line Profile

The line profile R_x at a frequency x observed in a lab frame is given by

$$R_x = \frac{\tilde{F}_x - \tilde{F}_c}{\tilde{F}_c}, \qquad (3.1.39)$$

where $\tilde{F}_x$ is the flux emergent from the star at the frequency x and $\tilde{F}_c$ is the flux in the continuum away from the line.

In the model we are working with, $s_0(p, x)$ defines a surface of constant velocity where s_0 is chosen such that

$$x = (s_0/r_0)V(r_0), \qquad (3.1.40)$$

where

$$r_0^2 = s_0^2 + p_0^2. \qquad (3.1.41)$$

r_0 denotes the value of r at the surface of constant radial velocity corresponding to x. From (3.1.18), writing $x' = x_c$ and the corresponding $r' = r_c$, we see that x_c is the value of x' at $r' = r_c$, and further

$$\mu' = \mu_c = \left[1 - (p/r_c)^2\right]^{1/2}. \qquad (3.1.42)$$

Then $\tilde{F}_x$ is defined by

$$\begin{aligned}\tilde{F}_x &= 2\pi \int_{-\infty}^{\infty} I(\infty, p, x) p\, dp \\ &= 2\pi \left[\int_{r_c}^{\infty} S(r_0) \left[1 - \exp(-\tau(-\infty, p, x))\right] p\, dp \right. \\ &\quad + \int_0^{r_c} S(r_0) \left\{1 - \exp(-\tau(-\infty, p, x))\Phi(x_c)\right\} p\, dp \\ &\quad \left. + \int_0^{r_c} I_c \exp\left\{-\tau(-\infty, p, x)\,\Phi(x_c)\right\} p\, dp \right]. \qquad (3.1.43)\end{aligned}$$

$\Phi(x)$ occurring in (3.1.43) is defined by (3.1.21) and (3.1.22). $\Phi(x_c)$ takes care of the effect of occultation of the medium by the core star. We recall that

$$\Phi(x_c) = \int_{-\infty}^{x_c} \phi(\xi')\, d\xi' \qquad (3.1.44)$$

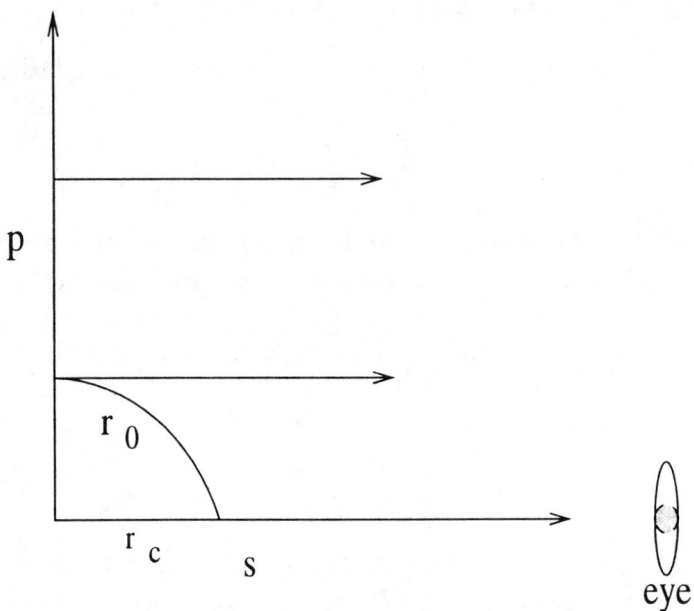

Figure 3.8. Direction of ray in spherical medium

$$\Phi(x_c) = 0 \quad \text{for } x' < 0 \\ = 1 \quad \text{for } x' > 0. \quad (3.1.45)$$

In (3.1.43), the first term accounts for the emission from the portion of the envelope outside the disc, $p > r_c$. In the second term, $\Phi(x_c)$ takes care of the correction for the occultation of the part of the envelope by the core. The third term gives the contribution to the flux from the radiation from the core.

Beyond the wings of the line, we have contribution to the flux from continuum radiation which is given by

$$\tilde{F}_c = 2\pi I_c \int_0^{r_c} p \, dp = \pi r_c^2 I_c. \quad (3.1.46)$$

Then from (3.1.39) and (3.1.43)–(3.1.46), we can determine R_x, the line profile at frequency x. Sometimes it is found convenient to change the interaction variable from p to r and express $\tilde{F}$, $\tilde{F}_c$ and R_x in terms of r.

Sobolev's theory is an elegant attempt at explaining the complex problem of line formation in moving media. He demonstrates that presence of significant velocity gradient in the medium contributes to the simplification of the problem. In the case of stationary medium, the rate of transition between any two levels $i \to j$ is compensated by the inverse transition $j \to i$. Further if the region is homogeneous, this balancing of forward and inverse transition can be taken for granted at every point as a fairly good approximation. However, in the presence of a velocity gradient in the medium this balance is disturbed. The rate of radiative transition in the upward direction falls short of that in the downward direction and the photons escape the medium. The difference in the rate of transition is given by

$$n_j A_{ji} \mathcal{P}_{ij},$$

where n_j is the atomic population at the jth level, A_{ji}, the relevant Einstein coefficient and $\mathcal{P}_{ij}$ is the escape probability of the photon due to Doppler shift of the profile. The scheme of obtaining $\mathcal{P}_{ij}$ by Sobolev's method for a given velocity field has been shown above. It was found that for complete redistribution of photons $\mathcal{P}_{ij}$ is independent of profile function. The determination of escape probability depends on the local values of integrated line opacity coefficient and the velocity gradient.

Mihalas (1978) considered a scheme of solution of multilevel problems in which one has to deal with a set of algebraic equations which is solved locally at each point of the medium. Doppler shift contributes to the destruction of interlocking of the radiation fields at different points. In the original work of Sobolev (1957), the line profiles were taken to be rectangular. In a subsequent work he showed that this restriction was not necessary. His method was a physical method where formal solution of radiative transfer equation was avoided. The emission line profiles and line intensities were calculated by inserting the escape probabilities into the equations of statistical equilibrium for sufficient number of energy levels. The use of Sobolev's method for solving multilevel radiative transfer problems is reviewed in Mihalas (1978) and the method in general has been extended by Kriz (1973) and Briot (1971).

It may be mentioned that in the escape probability method of Sobolev the velocity field in the medium can be represented by a succession of constant velocity surfaces labeled by different velocities. This takes account of the existence of velocity gradient in the atmosphere. When the flow speed far exceeds the mean thermal speed, an observer fixed in lab frame observes the radiation of each frequency come from specific velocity surface. The line of sight of the observer crosses each surface only once. Rybicki and Hummer (1978) drew attention to the fact that the projection of flow velocity on the line of sight may not always change monotonically in any direction. In that case, the line of sight might cross a constant velocity surface more than once. This is particularly the case for accelerating flows. Rybicki and Hummer (1978) used the basic tenets of Sobolev's theory to tackle this rather complex problem where one has to take care of the radiative coupling of two different points of the same velocity surface. The local character of the radiation field assumed in original reference of Sobolev is thus lost. Rybicki and Hummer (1978) solved this problem of non-local radiative coupling by what they called a generalised Sobolev method. This will be detailed in the following section.

3.2 Generalised Sobolev Method

Rybicki and Hummer (1978) assumed that the rays of radiation can intersect a particular velocity surface more than once. Their results dealt mainly with radiative transfer problems of spherical symmetry applicable to three-dimensional velocity field in the moving media. They used the relevant radiative transfer equations and their solutions. For simplicity, the time-independent transfer problems for only two-level atoms in a three-dimensional moving media will be considered.

The radiative transfer equation in this case is written as

$$\mathbf{n} \cdot \nabla I(\mathbf{r}, \mathbf{n}, \nu) = -\chi(\mathbf{r}) \, \phi[\nu - (\nu_0/c)\mathbf{n} \cdot \mathbf{V}(\mathbf{r})] \, [I(\mathbf{r}, \mathbf{n}, \nu) - S(\mathbf{r}, \nu)], \tag{3.2.1}$$

where $I(\mathbf{r}, \mathbf{n}, \nu)$ is the specific intensity at $\mathbf{r}$ in a direction $\mathbf{n}$ at a frequency ν and $\mathbf{V}(\mathbf{r})$ is the velocity field . The frequency integrated line opacity $\chi(r)$ for two-level atoms is given by

$$\chi(r) = (h\nu_0/4\pi)B_{12}n_l(r). \tag{3.2.2}$$

In (3.2.2), ν_0 is the line centre frequency, B_{12} the relevant Einstein coefficient, $n_l(r)$, population of the lower level. h and c are Planck constant and the velocity of light respectively.

For complete redistribution, the source function $S(\mathbf{r})$ is given by
$$S(\mathbf{r}) = [1 - \varepsilon(\mathbf{r})]\bar{J}(\mathbf{r}) + \varepsilon(\mathbf{r})B(\mathbf{r}), \qquad (3.2.3)$$
where $\varepsilon(\mathbf{r})$ is the ratio of collisional to total de-excitation from upper level, $B(\mathbf{r})$, the Planck function at $\mathbf{r}$ at the local electron temperature at frequency ν_0 at line centre.

The angle and frequency integrated mean intensity $\bar{J}(\mathbf{r})$ is given by
$$\bar{J}(\mathbf{r}) = \frac{1}{4\pi} \int d\Omega(\mathbf{n}) \int_0^\infty d\nu \, \phi[\nu - (\nu_0/c)\mathbf{n}\cdot\mathbf{V}(\mathbf{r})] \, I(\mathbf{r},\mathbf{n},\nu) \qquad (3.2.4)$$
where $\phi(\mathbf{r},\nu)$ is the profile function normalised as
$$\int_0^\infty \phi(\mathbf{r},\nu)d\nu = 1. \qquad (3.2.5)$$

Similarly, we can define $\bar{H}(\mathbf{r})$ as
$$\begin{aligned}\bar{H}(\mathbf{r}) &= \frac{1}{4\pi}\int d\Omega(\mathbf{n})\cdot\mathbf{n} \\ &\quad \times \int_0^\infty d\nu\,\phi[\nu - (\nu_0/c)\mathbf{n}\cdot\mathbf{V}(\mathbf{r})]\,I(\mathbf{r},\mathbf{n},\nu) \\ &= \frac{1}{4}\tilde{F}(\mathbf{r}).\end{aligned} \qquad (3.2.6)$$

Hence forward continuous absorption will be considered negligible: The formal solution of the radiative transfer equation (3.2.1) for intensity at a point $\mathbf{r}$ in a direction $\mathbf{n}$ at frequency ν obtained by integrating along the ray is given by
$$\begin{aligned}I(\mathbf{r},\mathbf{n},\nu) &= \int_0^{\bar{R}} S(\mathbf{r}-\mathbf{n}s)\exp\left\{-\int_0^s \chi(\mathbf{r}-\mathbf{n}s')\right. \\ &\quad \left.\times\phi\left[\nu-\frac{\nu_0}{c}\mathbf{n}\cdot\mathbf{V}(\mathbf{r}-\mathbf{n}s')\right]ds'\right\}\chi(\mathbf{r}-\mathbf{n}s)ds \\ &\quad + I_\nu^{inc}\exp\left\{-\int_0^{\bar{R}}\chi(\mathbf{r}-\mathbf{n}s')\right. \\ &\quad \left.\times\phi[\nu-(\nu_0/c)\mathbf{n}\cdot\mathbf{V}(\mathbf{r}-\mathbf{n}s')]ds'\right\}.\end{aligned} \qquad (3.2.7)$$

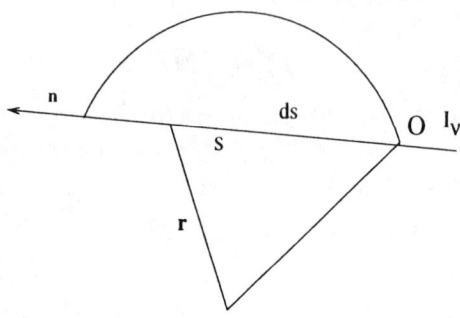

Figure 3.9. Direction of an ray through a spherical medium

$\overline{R}$ is the distance of the point of incidence of I_ν^{inc} which in the case of star is stellar surface.

Sobolev recognised that in the presence of large velocity gradient, the above equation is very much simplified due to favourable escape of the photons. We have seen that the source function S_x, the flux $\bar{F}_x$ and the line profile R_x can be calculated through the escape probability functions $\mathcal{P}_e$ and $\mathcal{P}_c$ in the Sobolev or supersonic approximation. R_x has been calculated for the velocity field $V = kr$ under the condition of complete redistribution of frequency. Once the opacity, the source function and line of sight velocity are known at a point, the intensity of radiation at that point can be calculated from (3.2.7).

Along a given ray, the specific intensity at a particular frequency ν remains the same except at specific points where the moving material has the Doppler shift just correct for permitting absorption and emission at frequency ν. These points occur where the following resonant condition holds good

$$(\nu - \nu_0)/\nu_0 = V_s/c, \qquad (3.2.8)$$

3.2 Generalised Sobolev Method

where V_s is the line of sight velocity $V_s = \mathbf{n} \cdot \mathbf{V}$. Rybicki and Hummer (1978) took note of the point that in the neigbourhood of resonant points χ and S vary slowly and can be taken outside the integral without encountering any serious error.

For monotonically increasing and monotonically decreasing laws of velocity, the field can be represented by a succession of constant velocity surfaces, the line of sight crossing each velocity surface only once. Sobolev's approximation is concerned with such single contact velocity surfaces only. The generalisation of Sobolev approximation by Rybicki and Hummer (1978) consists in extending this to situations where the line of sight crosses the same velocity surface more than once. The velocity field $V(r)$, in this case, may change sign in the medium. This results in the non-local interlocking of the radiative character of different points on the same velocity surface.

The salient features of the modification proposed by Rybicki and Hummer (1978) are as follows.

(a) They distinguish between the slowly varying and fast varying quantities occurring in the integral for intensity (3.2.7). The slowly varying χ and S are placed outside the integral in (3.2.7).

(b) The velocities are expanded in first order in s. In tensor notation

$$V_i(\mathbf{r} - \mathbf{n}s) = V_i(r) - \sum_j n_j \frac{\partial V_i}{\partial r_j} s$$

$$\Longrightarrow \mathbf{n} \cdot \mathbf{V}(\mathbf{r} - \mathbf{n}s) = \mathbf{n} \cdot \mathbf{V}(r) - Q(\mathbf{r} \cdot \mathbf{n})s$$

where

$$Q(\mathbf{r}, \mathbf{n}) = \sum_i \sum_j n_i n_j \frac{\partial V_i}{\partial V_j} = \frac{dV_s}{ds}. \qquad (3.2.9)$$

(c) Taking account of (a) and (b), equation (3.2.7) is reduced to

$$\begin{aligned}I(\mathbf{r}, \mathbf{n}, \nu) =\; & \chi S \int_0^{\overline{R}} ds\, \phi\left[\nu - (\nu_0/c)(\mathbf{n} \cdot \mathbf{V} + Qs)\right] \\& \times \exp\left\{-\chi \int_0^s ds'\phi\left[\nu - (\nu_0/c)(\mathbf{n} \cdot \mathbf{V} + Qs')\right]\right\} \\& + I_\nu^{inc} \exp\left\{-\chi \int_0^{\overline{R}} ds'\,\phi\left[\nu - \frac{\nu_0}{c}(\mathbf{n} \cdot \mathbf{V} + Qs')\right]\right\}.\end{aligned}$$
$$(3.2.10)$$

(d) The following substitutions are then made. We define

$$\lambda = \frac{s}{\Delta s} = \frac{s}{L}, \qquad (3.2.11)$$

$$\xi = \frac{\nu - (\nu_0/c)\mathbf{n}\cdot\mathbf{V}}{\Delta \nu}, \qquad (3.2.12)$$

where the line profile width $\Delta \nu$ can as well be taken as Doppler width. The normalisation condition can be written as

$$\int_{-\infty}^{\infty} \phi(\xi)\, d\xi = 1, \qquad (3.2.13)$$

where

$$\phi(\xi) = \Delta\nu(\xi \Delta\nu). \qquad (3.2.14)$$

L is the Sobolev length being the distance through which the profile function is Doppler shifted through its own width.

Using (3.1.8), (3.2.9) and (3.2.14) and writing the optical depth τ as

$$\tau = \chi \frac{\Delta s}{\Delta \nu} = \frac{\chi L}{\Delta \nu} = \frac{\chi c}{\nu_0 |Q|}, \qquad (3.2.15)$$

we have

$$I(\tau,\xi) = S\tau \int_0^{\overline{R}/L} d\lambda(\xi - \lambda)\, \exp\left\{-\tau \int_0^{\lambda} d\lambda'\, \Phi(\xi - \lambda')\right\}$$

$$+ I^{inc} \exp\left\{-\tau \int_0^{\overline{R}/L} d\lambda'\, \Phi(\xi - \lambda')\right\}. \qquad (3.2.16)$$

Since $\overline{R} \gg L$, the upper limits $\overline{R}/L$ in the integrals can be replaced by ∞. Writing

$$t = \xi - \lambda, \quad t' = \xi - \lambda'. \qquad (3.2.17)$$

and

$$w(t) = \int_{-\infty}^{t} \phi(t')\, dt' \qquad (3.2.18)$$

(3.2.10) is written as

$$I(\mathbf{r}, \mathbf{n}, \xi) = S\{1 - \exp[-\tau w(\xi)]\} + I^{inc} \exp[-\tau w(\xi)]. \qquad (3.2.19)$$

If $\xi \to -\infty$ specifies the direction of I_ν^{inc}, and if I_ν^{emg} denotes the intensity which escapes the resonant region until it escapes

3.2 Generalised Sobolev Method

the medium or meets another resonant region, we may treat I_ν^{emg} as the limit of I as $\xi \to +\infty$. Then equation (3.2.19) leads to

$$I_\nu^{emg} = I_\nu^{inc} e^{-\tau} + S(1 - e^{-\tau}). \tag{3.2.20}$$

When the photons escaping the resonant region meets the second resonant region, I_ν^{emg} acts as the incident intensity for the second resonant region. The frequency integrated intensity can be obtained from (3.2.19) as

$$\overline{I}(\mathbf{r}, \mathbf{n}) = \int_{-\infty}^{\infty} \phi(\xi) \left\{ S[1 - e^{-\tau\omega(\xi)}] + I_\nu^{inc} e^{-\tau\omega(\xi)} \right\} d\xi. \tag{3.2.21}$$

Now changing the variable to $\omega(\xi)$ and using (3.2.18) we have

$$\overline{I}(\mathbf{r}, \mathbf{n}) = S(\mathbf{r}) \left[1 - \frac{1 - e^{-\tau}}{\tau} \right] + I_\nu^{inc} \frac{1 - e^{-\tau}}{\tau} \tag{3.2.22}$$

where the incident radiation I^{inc} is evaluated at

$$\overline{\nu} = \nu_0 + (\nu_0/c)\mathbf{n} \cdot \mathbf{V}(r). \tag{3.2.23}$$

Equations (3.2.19), (3.2.22) and (3.2.23) are the basic equations of Sobolev's method when the line-of-sight crosses any constant velocity surface only once.

We have outlined above the Rybicki and Hummer's (1978) scheme of generalised Sobolev method for monotonically increasing or decreasing velocity law. However, they also demonstrated that above equations can be adopted to the problems of multiple surfaces. For example, for problems of double surfaces, we may write for the intensity passing through the two surfaces by applying (3.2.19) to each surface and taking note of the fact that the intensity emergent from the first surface will be the incident intensity for the second surface.

Then the emergent intensity after passing through the double surfaces, can be written as

$$I_\nu^{emg} = e^{-\tau_1} e^{-\tau_2} I_\nu^{inc} + e^{-\tau_1}(1 - e^{-\tau_2})S_2 + (1 - e^{-\tau_1})S_1 \tag{3.2.24}$$

where τ_1 and S_1 refer to the first surface and τ_2 and S_2 to the second surface.

The frequency integrated intensity $\overline{I}(\mathbf{r}, \mathbf{n})$ can be written as

$$\bar{I}(\mathbf{r},\mathbf{n}) = e^{-\tau_2}\left(\frac{1-e^{-\tau_1}}{\tau_1}\right)I_{\bar{\nu}}^{inc} + (1-e^{-\tau_2})\left(\frac{1-e^{-\tau_1}}{\tau_1}\right)S_2(\mathbf{r})$$
$$+ \left[1 - \left(\frac{1-e^{-\tau_1}}{\tau_1}\right)\right]S_1(\mathbf{r}). \qquad (3.2.25)$$

For the details of the use of generalised Sobolev method and its various applications the readers are referred to the paper of Rybicki and Hummer (1978), Marti and Noerdlinger (1972) and Grachev and Grinin (1975).

Sobolev's escape probability method was originally proposed to explain the problem of line formation in extensive moving media. It is sometimes termed as "first-order escape probability method" to distinguish it from other probabilistic methods proposed by Athay (1972), Frisch and Frisch (1975), Canfield (1981) and Scharmer (1981, 1984). The latter mentioned probabilistic methods are called "second order probability methods". The aim of the second order method is to remove some of the limitations of the first order one. One such limitation is that it works well with line formation problems with complete redistribution of frequency but not with those involving partial redistribution. The second order methods aim at building operators for solving the relevant system of equations which are economical on computer time. The model of the atmosphere is usually taken to be plane parallel and static and occasionally spherical and static. The core saturation method of Rybicki (1984), operator perturbation method of Cannon (1973) and Kalkofen (1987) and Scharmer's (1981, 1984) formulation of line transfer problem have as their basis the escape probability of photons from the participating media. The two treatise of Kalkofen (1984, 1987) contain excellent reviews of these methods. The features of these methods relevant to the solution of transfer problems in spherically symmetric moving media will be taken up in chapter 5 and chapter 8 of this book.

3.3 Methods Based on Rosseland Cycle and on Diffusion of Free Electrons

Methods Based on Rosseland Cycle

This method was developed by Miyamoto (1949–1952), Kogure (1949–1967) and their coworkers. The method was diagnostic in nature aiming at explaining the abnormal line characteristics of Be-stars in particular. The basic physical characteristic of Be-star atmosphere is the existence of rapid rotation in it about an axis passing through the centre of the star. Miyamoto *et al.* tried to explain the radiation field and the dynamical properties of Be-stars in terms of a "dilution factor". They did not consider the effect of differential Doppler effect on the transfer of radiation in the stellar envelope. Their method is a static method built on the earlier ideas of Rosseland (1926), Zanstra (1934) and Heyney (1938).

The physical model of the atmosphere of Be-stars was conceived to be a relatively cool envelope surrounding a central star exposed to ultraviolet rays emanating from the star. In the visible region, the stellar envelope was taken to be optically thin for continuous radiation but optically thick for Balmer lines of hydrogen spectra. The second property was assumed to explain the appearance of strong shell absorption lines in Be-shell stars. Balmer decrement observed varied from star to star and time to time.

The physical properties of Be-envelopes are supposed to stand somewhere between normal stellar photosphere (dilution factor of the order of unity) and the planetary nebular shell (dilution factor 10^{-1} to 10^{-13}). The dilution factor of Be-envelope lies between 10^{-1} and 10^{-3}. Furthermore, in the Be-envelopes the photoionisation seems to take place from the second energy level of hydrogen atom instead of ground level. In that sense Be-envelope does not conform fully to that of the Rosseland cycle. The effective adherence of a system to Rosseland cycle suggests that the ultraviolet radiation from the core stars induces photoionisation from the ground state of hydrogen followed by recombination and cascade transition producing emission line spectra. This condition prevails in the planetary nebular shells. Thus the fact that the photoionisation in the Be-envelope takes place from second level of hydrogen instead of ground level suggests that Rosseland cycle is incomplete in Be-atmospheres.

In the static approach put forward by Miyamoto (1949), Kogure (1959) et al., the facts stated above were taken into consideration. The equations of radiative transfer and statistical equilibrium were solved for lower order atomic level populations without taking account of the existence of velocity gradient. The stellar envelope was divided into an appropriate number of equal velocity zones for the calculation of line profiles of Be-star. The three-level problem of hydrogen atoms was first solved by Miyamoto (1949) and four to seven levels problems by Miyamoto (1952), Kogure (1959, 1961, 1967) and Potasch (1961). Three discrete energy levels were taken to be averaged continuum, ground level and the first excited level. The cyclic equations for these three levels and transfer equations for Lyman-continuum $(1-c)$ and Lyman $\alpha(1-2)$ radiation were simultaneously solved. It was found that in the first approximation, the Rosseland cycle $(1 \to c \to 2 \to 1)$ was balanced by its inverse. Under the circumstance, the ionisation from the second level assumed importance in the study of spectrum of the Be-stars. Miyamoto extended the scheme to solve the problems with four and five energy levels and compared the Balmer decrements $H_\alpha : H_\beta$ in each case with the observations. In Miyamoto's calculations the atmospheres of Be-stars were assumed to be sufficiently opaque to the radiation from subordinate levels. However, Kogure in some cases took the radiation from subordinate level to be transparent beyond the Paschen series. The population of the second energy level was taken to be the same for three to seven energy level calculations.

Thus the scheme of Kogure differed from that of Miyamoto in the choice of optical thickness of the envelope. Potasch (1961) included the contribution of collisional transitions. Their main object of study was the variation of Balmer decrement H_α/H_β with the variation of dilution factor W. The Balmer decrement was found to depend on optical thickness and collision effects. In the above methods agreement was sought between the spectral structure suggested by static microscopic theory of transfer of radiation through *Be*-star envelopes and the observed features of spectral lines. This agreement between the theory and observation was secured in each case by adjusting the physical properties of the stellar atmosphere and the geometrical dilution factor.

Methods Based on Diffusion by Free Electrons

In the early years attempt was also made to explain the appearance of emission lines in the spectra of certain classes of stars to be due to the diffusion of radiation by the free electrons in their envelopes. Schuster (1905) in examining the transfer of radiation in a static, dense, scattering atmosphere found that emission components might appear in special spectra due to pushing down of the continuum around the position of the line. He concluded that wide variety of spectral lines such as purely absorption, purely emission or the combination of them could be obtained by studying the relation between the gradient of thermal emission and the coefficient of diffusion. Schuster concluded that the emission component appeared when the gradient of temperature in the medium is small.

Underhill (1949) drew attention to the importance of electron scattering in explaining the depression of continuum suggested by Schuster. Burbridge (1952) studied the limitation of this approach by fairly rigorous mathematical reasoning.

References

Athay, R. (1972): Radiative transport in spectral lines. Dordrecht,Ridel.
Athay, R. (1931): Ap. J., **176**, 659.
Beals, C. (1929): MNRAS, **90**, 966.
Beals, C.(1930): Publ. Dom. Astrophy.Obs. Victoria, **4**, 271.
Beals, C. (1934): Publ. Dom. Astrophy.Obs. Victoria, **6**, 95.
Beals, C. (1950): Publ. Dom. Astrophy.Obs. Victoria, **9**, 1.
Briot, D. (1971): Astr. Ap., **11**, 57.
Briot, D. (1977): Astr. Ap., **54**, 599.
Briot, D. (1981a): Astr. Ap., **103**, 5.
Briot, D. (1981b): Astr. Ap., **103**, 1.
Burbridge, G.R., Burbridge, E.M. (1953): Ap. J., **118**, 252.
Burbridge, G.R., Burbridge, E.M., (1954): Ap. J., **119**, 496.
Canfield, R.C., Puetter, R.C. (1981): Ap. J., **243**, 381.
Canfield, R.C., Puetter, R.C. Ricchiazzi,P.J. (1981): Ap. J., **248**, 82.

Cannon, C.J. (1973): JQRST, **13**, 627.
Frisch, U. Frisch, H. (1975): MNRAS, **173**, 167.
Grachev, S.L. Grinin, V.P. (1975): Astrophysics, **11**, 20.
Heyney, L.G. (1938):Ap. J., **88**, 133.
Kalkofen, W. (1984): Methods in Radiative Transfer. Cambridge University Press, Cambridge.
Kalkofen, W. (1987): Numerical Radiative Transfer. Cambridge University Press, Cambridge.
Kogure, T. (1959): Publ. Astro. Soc. Japan, **11**, 127.
Kogure, T. (1959): Publ. Astro. Soc. Japan, **11**, 278.
Kogure, T. (1961): Publ. Astro. Soc. Japan, **13**, 335.
Kogure, T. (1967): Publ. Astro. Soc. Japan, **19**, 30.
Kriz, S. (1973): Bull. Astro. Inst. Czech, **25**, 143.
Kriz, S. (1976): Bull. Astro. Inst. Czech, **27**, 321.
Kriz, S. (1979): Bull. Astro. Inst. Czech, **30**, 83.
Kriz, S. (1979): Bull. Astro. Inst. Czech, **30**, 93.
Marti, F., Noerdlinger, P.D. (1977): Ap. J.,**215**, 247.
Mihalas, D. (1978): Stellar Atmospheres. W. Freeman and Co., San Francisco.
Miyamoto, S. (1949): Jap. J. Astron., **1**, 17.
Miyamoto, S. (1952): Publ. Astro. Soc. Japan., **4**, 1.
Miyamoto, S. (1952): Publ. Astro. Soc. Japan., **4**, 28.
Münch, G. (1950): Ap. J., **112**, 266.
Potasch, S.R. (1961): Ann. d'Ap., **24**, 159.
Rosseland, S. (1926): Ap. J., **63**, 218.
Rybicki, G.B., Hummer, D.G. (1978): Ap. J., **219**, 654.
Rybicki, G.B. (1984): Escape probability methods. Methods in Radiative Transfer, ed. Kalkofen, W. Cambridge University Press, Cambridge, p. 21.
Scharmer, G.B. (1981): Ap. J., **249**, 720.
Scharmer, G.B. (1984): Escape probability methods. Methods in Radiative Transfer.ed. Kalkofen, W. Cambridge University Press, Cambridge, p 21.
Schuster, A. (1905): Ap. J., **21**, 1.
Sobolev, V.V. (1957): Sov. Astron., **1**, 678
Sobolev, V.V. (1960): Moving Atmospheres of Stars. Harvard University Press, Cambridge.
Thomas, R.N. (1949): Ap. J., **109**, 500.
Underhill, A.B. (1949): MNRAS, **109**, 562.
Underhill, A.B. (1949): MNRAS, **110**, 340.
Zanstra, H. (1941): MNRAS, **101**, 273.

Chapter 4. Impact Parameter Method for Transfer Equation in Lab Frame

In this chapter, we shall review the impact parameter method of Feautrier (1964), which has proved to be an effective, versatile and stable computational method for solving radiative transfer problems in plane parallel medium. Subsequently it has been found that with some small modifications the method can be successfully used to solve radiative transfer problems in spherically symmetric stationary and moving media.

The essential feature of the method consists of effecting a change of variable in transfer equation from specific intensity to mean intensity and flux-like variable expressed in terms of two intensities in opposite directions. By clever manipulation of these two variables, Feautrier converts the first order integro-differential equation of transfer to a second order differential equation and then solves the resulting two-point boundary value problem. The transfer equation and the boundary conditions are first written as finite difference equations by discretisation of depth, angle and frequency parameters and the other variables are expressed in terms of them. The transfer equation and boundary conditions are written into block tridiagonal system and solved by Gaussian elimination scheme.

For completeness we shall first outline Feautrier's method as used for the solution of radiative transfer problems in slab medium and spherically symmetric stationary medium. This will allow us to understand the basics of the method and trace its evolution leading to the solution of radiative transfer problems in spherically symmetric moving media.

4.1 Impact Parameter Method in Plane Geometry for Stationary Media

The time-independent transfer equation for a static plane parallel medium can be written in terms of two oppositely directed pencils $\pm\mu$ as

$$\pm\mu[\partial I(z,\mu,\nu)/\partial z] = \chi(z,\nu)[S(z,\nu) - I(z,\mu,\nu)] \qquad (4.1.1)$$

where the symbols have their usual meanings [cf. equations (1.2.17)–(1.2.21) and Fig. 1.2].

We define u and v, the mean intensity and flux like variables as

$$\left. \begin{array}{l} u = \frac{1}{2}[I^+(z,\mu,\nu) + I^-(z,\mu,\nu)] \\ v = \frac{1}{2}[I^+(z,\mu,\nu) - I^-(z,\mu,\nu)] \end{array} \right\} \qquad (4.1.2)$$

where

$$\left. \begin{array}{l} I^+(z,\mu,\nu) = I(z,+\mu,\nu) \\ I^-(z,\mu,\nu) = I(z,-\mu,\nu) \end{array} \right\} \qquad (4.1.3)$$

with $0 \leq \mu \leq 1$.

Adding and subtracting the two equations in (4.1.1), we have

$$\mu \frac{\partial v(z,\mu,\nu)}{\partial z} = \chi(z,\nu)[S(z,\nu) - u(z,\mu,\nu)], \qquad (4.1.4)$$

and

$$\mu \frac{\partial u(z,\mu,\nu)}{\partial z} = -\chi(z,\nu)v(z,\mu,\nu). \qquad (4.1.5)$$

Eliminating v from (4.1.4) and (4.1.5), we have

$$\mu^2 \frac{\partial^2 u(\tau_\nu,\mu,\nu)}{\partial \tau_\nu^2} = u(\tau_\nu,\mu,\nu) - S(\tau_\nu,\nu), \qquad (4.1.6)$$

where $d\tau_\nu = -\chi_\nu\, dz$.

A unique solution of (4.1.6) can be obtained by specifying the boundary conditions at $\tau_\nu = 0$ and $\tau_{\max}$, labelling respectively the upper and lower boundary of the medium under consideration.

A sample of boundary conditions common to stellar atmosphere problems are:

(a) at the upper boundary of the medium denoted by

4.1 Impact Parameter Method in Plane Geometry for Stationary Media

$$\tau_\nu = 0, \quad I^-(0, \mu, \nu) = 0.$$

This implies that $v(\tau_\nu = 0) = u(\tau_\nu = 0)$. Then

$$\mu \left| \frac{\partial u(\tau_\nu, \mu, \nu)}{\partial \tau_\nu} \right|_{\tau_\nu = 0} = u(0, \mu, \nu), \qquad (4.1.7)$$

and

(b) at the lower bounding surface

$$\tau_\nu = \tau_{\max}, \quad I(\tau_{\max}, \mu, \nu) = I^+(\tau_{\max}, \mu, \nu).$$

This implies $v(\tau_{\max}) = I^+(\tau_{\max}, \mu, \nu) - u(\tau_{\max})$, and it yields

$$\mu \left| \frac{\partial u(\tau_\nu, \mu, \nu)}{\partial \tau_\nu} \right|_{\tau_{\max}} = I^+(\tau_{\max}) - u(\tau_{\max}). \qquad (4.1.8)$$

In semi-infinite medium, the diffusion approximation may be invoked at $\tau_{\max}$. In this case,

$$I(\tau_{\max}, \mu, \nu) = B_\nu(\tau_{\max}) + \mu \left. \frac{\partial B_\nu}{\partial \tau_\nu} \right|_{\tau_{\max}}, \qquad (4.1.9)$$

from where, we have

$$\left. \begin{array}{rl} u(\tau_{\max}) = & B_\nu(\tau_{\max}) \\ v(\tau_{\max}) = & \mu \left| \frac{\partial B_\nu}{\partial \tau_\nu} \right|_{\tau_{\max}} \end{array} \right\} \qquad (4.1.10)$$

Discretisation of Variables and Difference Equation Representation

The depth, angle and frequency variables are discretised by choosing their values at conveniently chosen discrete set of points. The second order transfer equation (4.1.6) and the boundary conditions are expressed as finite difference equations.

The process of discretisation is done as follows.

$$\begin{array}{rl} \text{Depth points} : & \{\tau_d\}, \, (d = 1, 2, \ldots D), \tau_1 < \tau_2 < \cdots < \tau_D, \\ \text{Angle points} : & \{\mu_i\}, \, (i = 1, 2, \ldots I), 0 \leq \mu \leq 1, \\ \text{Frequency points} : & \{\nu_n\}, \, (n = 1, 2, \ldots N), 0 < \nu < \infty. \end{array} \qquad (4.1.11)$$

Any variable represented by

$$X_{din} \equiv X(\tau_d, \mu_i, \nu_n) \qquad (4.1.12)$$

is evaluated on the shell surface d. The cells $d - (d-1)$ may be of unequal size. The representation $X_{(d+\frac{1}{2}),i,n} \equiv X(\tau_{d+\frac{1}{2}}, \mu_i, \nu_n)$ imply that the variables have their values specified at the midpoint of the space confined between d and $(d+1)$. The integrals involving angle and frequency, whenever they occur are replaced by quadrature. The combinations of angles and frequencies are grouped into a single series with index $k = 1, 2, \ldots, K$ with $K = IN$. $(\mu_k, \nu_k) \to (\mu_i, \nu_n)$ where $k = i + (n-1)I$. Now replacing the integrals by quadrature sums (cf. 1.2.11) the source function $S_{(d+\frac{1}{2}),n}$ can be written as

$$S_{(d+\frac{1}{2}),n} = \alpha_{(d+\frac{1}{2}),n} \sum_{k=1}^{K} w_k\, u_{(d+\frac{1}{2}),k} + \beta_{(d+\frac{1}{2}),n} \qquad (4.1.13)$$

where $\alpha_{(d+\frac{1}{2}),n}$ is connected with the scattering coefficients. The first term in (4.1.13) contains the scattering integral and the second the thermal term. w_k is the appropriate weight for the combined angle frequency quadrature (which may contain line profile function as well).

Thus the finite difference equation representation of (4.1.6), the second order radiative transfer equation is

$$\frac{1}{\Delta\tau_{(d+\frac{1}{2}),k}} \left[\left(\frac{\mu_k^2}{\Delta\tau_{(d+1),k}} \right) u_{(d+\frac{3}{2}),k} \right.$$
$$\left. - \mu_k^2 \left(\frac{1}{\Delta\tau_{d,k}} + \frac{1}{\Delta\tau_{(d+1),k}} \right) u_{(d+\frac{1}{2}),k} + \left(\frac{\mu_k^2}{\Delta\tau_{d,k}} \right) u_{(d-\frac{1}{2}),k} \right]$$
$$\equiv u_{(d+\frac{1}{2}),k} - S_{(d+\frac{1}{2}),k} \qquad (4.1.14)$$

with $k = (1, 2, \ldots, K)$, $d = [2, 3, \ldots, (D-1)]$. In (4.1.14)

$$\Delta\tau_{d,k} = \frac{1}{2}\left[\chi_{(d-\frac{1}{2}),k}\Delta z_{(d-\frac{1}{2})} + \chi_{(d+\frac{1}{2}),k}\Delta z_{(d+\frac{1}{2})}\right], \quad (4.1.15)$$

$$\Delta\tau_{(d+\frac{1}{2}),k} = \frac{1}{2}\left[\Delta\tau_{d,k} + \Delta\tau_{(d+1),k}\right], \quad (4.1.16)$$

$$\left(\frac{du}{d\tau}\right)_d = \left[u_{d+\frac{1}{2}} - u_{d-\frac{1}{2}}\right]/\Delta\tau_d, \quad (4.1.17)$$

and

$$\left(\frac{d^2u}{d\tau^2}\right)_{d+\frac{1}{2}} = \left[\left(\frac{du}{d\tau}\right)_{d+1} - \left(\frac{du}{d\tau}\right)_d\right]/\Delta\tau_{(d+\frac{1}{2})}. \quad (4.1.18)$$

The source function $S_{(d+\frac{1}{2}),k}$ is given by (4.1.13). At each of the $D-2$ depth points, there is one equation for each value of k.

The boundary conditions (4.1.7) and (4.1.8) are now expressed as finite difference equations.

The boundary condition at upper surface can be stated as finite difference equation of (4.1.7). The boundary condition (4.1.7) implies $v(\tau_\nu = 0) = u(\tau_\nu = 0)$. From (4.1.4), we have the finite difference equation as

$$\mu_k[(v_{2,k} - v_{1,k})/\Delta\tau_{3/2,k}] = u_{3/2,k} - S_{3/2,k},$$

which implies

$$v_{2,k} = v_{1,k} + (\Delta\tau_{3/2,k}/\mu_k)[u_{3/2,k} - S_{3/2,k}]. \quad (4.1.19)$$

From (4.1.5), we have

$$v_{2,k} = \mu_k[u_{5/2,k} - u_{3/2,k}]/\Delta\tau_{2,k}. \quad (4.1.20)$$

From (4.1.7),

$$u_{1,k} = v_{1,k} = \mu_k[u_{3/2,k} - u_{1,k}]/\frac{1}{2}\Delta\tau_{3/2,k}. \quad (4.1.21)$$

Hence

$$u_{1,k} = v_{1,k} = u_{3/2,k}\bigg/\left[1 + \frac{1}{2\mu_k}\Delta\tau_{3/2,k}\right]. \quad (4.1.22)$$

Then substituting (4.1.20) and (4.1.21) in (4.1.19), we have for the difference equation for the boundary condition at upper surface as

$$\mu_k[u_{5/2,k} - u_{3/2,k}]/\Delta\tau_{2,k} = u_{3/2,k}/[1 + \frac{1}{2\mu_k}\Delta\tau_{3/2,k}]$$
$$+[\Delta\tau_{3/2,k}/\mu_k][u_{3/2,k} - S_{3/2,k}]. \quad (4.1.23)$$

Similarly it can be shown that the finite difference equation for the boundary condition (4.1.8) at the lower boundary is given by

$$\mu_k\left[u_{(D+\frac{1}{2}),k} - u_{(D-\frac{1}{2}),k}\right]/\Delta\tau_{D,k} = I^+_{(D+\frac{1}{2}),k} - u_{(D+\frac{1}{2}),k}$$
$$- \left(\frac{1}{2\mu_k}\Delta_{D,k}\right)$$
$$\times \left[u_{(D+\frac{1}{2}),k} - S_{(D+\frac{1}{2}),k}\right]. \quad (4.1.24)$$

Feautrier's Method of Solution

This method was originally developed to calculate the radiation field in plane parallel medium with the source function explicitly involving contribution from scattering and thermal sources.

The following two vectors are defined.

$$\mathbf{U}_{(d+\frac{1}{2})} \equiv \left[u_{(d+\frac{1}{2}),1}, u_{(d+\frac{1}{2}),2}, u_{(d+\frac{1}{2}),3}, \ldots, u_{(d+\frac{1}{2}),k}\right]. \quad (4.1.25)$$

At a particular depth $(d + \frac{1}{2})$, the vector contains all the components of angle and frequency. Secondly

$$\mathbf{L}_{(d+\frac{1}{2})} \equiv \left[\beta_{(d+\frac{1}{2}),1}, \beta_{(d+\frac{1}{2}),2}, \ldots, \beta_{(d+\frac{1}{2}),k}\right]. \quad (4.1.26)$$

$\mathbf{L}$ contains angle frequency components of the thermal part of source function.

With the above discretisation, the transfer equation (4.1.14) and the boundary conditions (4.1.23) and (4.1.24) can be combined to be written as a block tridiagonal system of the form

$$-\mathbb{A}_{(d+\frac{1}{2})}\mathbf{U}_{(d-\frac{1}{2})} + \mathbb{B}_{(d+\frac{1}{2})}\mathbf{U}_{(d+\frac{1}{2})} - \mathbb{C}_{(d+\frac{1}{2})}\mathbf{U}_{(d+\frac{3}{2})} = \mathbf{L}_{(d+\frac{1}{2})}. \quad (4.1.27)$$

The dimension of each one of the matrices $\mathbb{A}_{(d+\frac{1}{2})}$, $\mathbb{B}_{(d+\frac{1}{2})}$ and $\mathbb{C}_{(d+\frac{1}{2})}$ is $K \times K$. In this $\mathbb{B}_{(d+\frac{1}{2})}$ is a full matrix while $\mathbb{A}_{(d+\frac{1}{2})}$ and $\mathbb{C}_{(d+\frac{1}{2})}$ are diagonal matrices. $\mathbb{A}_{(d+\frac{1}{2})}$ and $\mathbb{C}_{(d+\frac{1}{2})}$ contain the

4.1 Impact Parameter Method in Plane Geometry for Stationary Media

finite difference representation of the differential operator for all angle-frequency points down the diagonal. $\mathbb{B}_{(d+\frac{1}{2})}$ contains the finite difference operators down the diagonal plus the diagonal and off-diagonal terms arising from the quadrature representation of the scattering integral in the source function. The vector $\mathbf{L}_{(d+\frac{1}{2})}$ contains the thermal terms in the source function. It is to be noted that the upper boundary condition suggests that $\mathbb{A}_{3/2} = 0$ and the lower boundary condition that $\mathbb{C}_{(D+\frac{1}{2})} = 0$.

Elimination Scheme

The Gaussian elimination scheme is used for solving the equation (4.1.27). At level $(d+\frac{1}{2})$, $\mathbf{U}_{(d+\frac{1}{2})}$ is expressed in terms of $\mathbf{U}_{(d+\frac{3}{2})}$ and this relation is used to eliminate $\mathbf{U}_{(d+\frac{1}{2})}$ from the next equation. We start the elimination scheme from the upper boundary where $\mathbb{A}_{3/2} = 0$.

Hence

$$\begin{aligned}\mathbf{U}_{3/2} &= (\mathbb{B}_{3/2}^{-1}\,\mathbb{C}_{3/2})\,\mathbf{U}_{5/2} + (\mathbb{B}_{3/2}^{-1}\,\mathbf{L}_{3/2}) \\ &= \mathbb{D}_{3/2}\,\mathbf{U}_{5/2} + \mathbf{V}_{3/2}.\end{aligned} \qquad (4.1.28)$$

We substitute (4.1.28) in (4.1.27) and obtain for $d = 2$.

$$\begin{aligned}\mathbf{U}_{5/2} &= (\mathbb{B}_{5/2} - \mathbb{A}_{5/2}\,\mathbb{D}_{3/2})^{-1}\,\mathbb{C}_{5/2}\,\mathbf{U}_{7/2} \\ &\quad + (\mathbb{B}_{5/2} - \mathbb{A}_{5/2}\,\mathbb{D}_{3/2})^{-1}\,[\mathbf{L}_{5/2} + \mathbb{A}_{5/2}\,\mathbf{V}_{3/2}] \\ &= \mathbb{D}_{5/2}\,\mathbf{U}_{7/2} + \mathbf{V}_{5/2}.\end{aligned} \qquad (4.1.29)$$

Proceeding in the same way, in general, we can write

$$\mathbf{U}_{(d+\frac{1}{2})} = \mathbb{D}_{(d+\frac{1}{2})}\mathbf{U}_{(d+\frac{3}{2})} + \mathbf{V}_{d+\frac{1}{2}} \qquad (4.1.30)$$

with

$$\mathbb{D}_{(d+\frac{1}{2})} = [\mathbb{B}_{(d+\frac{1}{2})} - \mathbb{A}_{(d+\frac{1}{2})}\,\mathbb{D}_{(d-\frac{1}{2})}]^{-1}\,\mathbb{C}_{(d+\frac{1}{2})} \qquad (4.1.31)$$

and

$$\begin{aligned}\mathbf{V}_{(d+\frac{1}{2})} &= [\mathbb{B}_{(d+\frac{1}{2})} - \mathbb{A}_{(d+\frac{1}{2})}\mathbb{D}_{(d-\frac{1}{2})}]^{-1}[\mathbf{L}_{(d+\frac{1}{2})} \\ &\quad + \mathbb{A}_{(d+\frac{1}{2})}\mathbf{V}_{(d-\frac{1}{2})}].\end{aligned} \qquad (4.1.32)$$

4. Impact Parameter Method for Transfer Equation in Lab Frame

To compute $\mathbb{D}_{(d+\frac{1}{2})}$ and $\mathbf{V}_{(d+\frac{1}{2})}$, we start from the lower boundary where $\mathbb{C}_{(D+\frac{1}{2})} = 0$. Then from (4.1.31), we have $\mathbb{D}_{(D+\frac{1}{2})} = 0$ and from (4.1.30)

$$\mathbf{U}_{(D+\frac{1}{2})} = \mathbf{V}_{(D+\frac{1}{2})}. \tag{4.1.33}$$

If we know $\mathbf{U}_{(D+\frac{1}{2})}$ we can obtain $\mathbf{U}$ for $d = (D-1), (D-2), (D-3), \ldots, 1$ by successive back substitution. Thus making use of upper and lower boundary conditions and by forward and backward substitution one can evaluate $\mathbf{U}_{(d+\frac{1}{2}),n}$ and $S_{(d+\frac{1}{2}),n}$ using suitable angle quadrature.

Feautrier's method is compact, elegant and stable for computation. The computing time for this method can be estimated by ascertaining the number of multiplications involved in solving the problem. For linear system of order n, the number of operations involved is of $O(n^3)$. In Feautrier's method it is proportional to DK^3 or DI^3N^3 where I is the number of angle points and N, the number of frequency points. Usually I is not very large but N can be large in the case of non-coherent scattering. For coherent scattering $N = 1$ and Feautrier's method is fast and economical in this case.

In many static media, the scattering of radiation is isotropic. In such cases angular mesh can be avoided by taking $\mathbf{J}_\nu$ in place of $\mathbf{U}_{\mu\nu}$. Instead of transfer equations, we use the moment equations and use the variable Eddington factor f_ν to state the closure conditions. This alternate scheme simplifies the method of solution of problems in the case of isotropically scattering media. In this method, we integrate (4.1.6) and the boundary conditions (4.1.7) and (4.1.8) over angles and get

$$\frac{\partial^2 (f_\nu J_\nu)}{\partial \tau_\nu^2} = J_\nu - S_\nu, \tag{4.1.34}$$

$$\left[\frac{\partial (f_\nu J_\nu)}{\partial \tau_\nu}\right]_{\tau_\nu=0} = H_\nu(\tau_{\nu=0}), \tag{4.1.35}$$

4.1 Impact Parameter Method in Plane Geometry for Stationary Media

and

$$\left[\frac{\partial(f_\nu J_\nu)}{\partial \tau_\nu}\right]_{\tau_\nu = \tau_{\max}} = (I^+)_{\tau_{\max}} - (f_\nu J_\nu)_{\tau_{\max}}, \qquad (4.1.36)$$

where f_ν, the variable Eddington factor is given by $f_\nu = K_\nu/J_\nu$.

Incidentally it may be mentioned that in the case of semi-infinite media, the diffusion approximation gives the lower boundary condition. This, in the moment form, reads as

$$\left[\frac{\partial(f_\nu J_\nu)}{\partial \tau_\nu}\right]_{\tau_{\max}} = \frac{1}{3}\left(\frac{1}{\chi_\nu}\left|\frac{\partial B_z}{\partial z}\right|\right)_{\tau_{\max}}. \qquad (4.1.37)$$

The equations (4.1.34), (4.1.35) and (4.1.36) or (4.1.37) are expressed as finite-difference equations as demonstrated in the Feautrier's angle-dependent equations and they lead to a block tridiagonal system of type (4.1.27). They can be solved by the Gaussian elimination scheme as in the earlier case.

However, to solve the finite-difference equations corresponding to (4.1.34)–(4.1.36), the depth variation of f_ν and $H_\nu(\tau_\nu = 0)$ must be known. To find them, an iterative scheme is followed with the starting point $S_\nu = B_\nu$, where B_ν is the Planck function whose temperature structure is known. This source function being given we can find the solution of transfer equation. For this depth variation at a specific frequency and angle (μ_i, ν_n), u is represented by a column vector $\mathbf{U}_k$ and the source function S by a column vector $\mathbf{S}_k$ which are given by

$$\mathbf{U}_k = [u_{(3/2),k}, u_{(5/2),k}, \ldots, u_{(D+\frac{1}{2}),k}] \qquad (4.1.38)$$

and

$$\mathbf{S}_k = [S_{(3/2),k}, S_{(5/2),k}, \ldots, S_{(D+\frac{1}{2}),k}]. \qquad (4.1.39)$$

Then the transfer equation (4.1.14) and the boundary conditions (4.1.23) and (4.1.24) are grouped together to give a tridiagonal matrix equation of the form

$$\mathbb{T}_k \mathbf{U}_k = \mathbf{S}_k, \qquad (4.1.40)$$

where $\mathbb{T}_k$ is a tridiagonal matrix of dimension $D \times D$. The equation (4.1.40) can be solved by Gaussian elimination scheme which has been elaborated in (4.1.28)–(4.1.33) and the paragraph preceding and following them. $\mathbf{U}_{(d+\frac{1}{2}),k}$ for $d = (1, 2, \ldots, D)$ and $k = (1, 2, \ldots, K)$ are determined and the mean intensity $J_{(d+\frac{1}{2}),k}$, $H_{(d+\frac{1}{2}),k}$ and $K_{(d+\frac{1}{2}),k}$ and hence $f_{(d+\frac{1}{2}),k}$ can be determined. Thus

$$J_{(d+\frac{1}{2}),k} = \sum_i w_i\, u_{(d+\frac{1}{2}),i,k} \qquad (4.1.41)$$

$$H_{(d+\frac{1}{2}),k} = \sum_i w_i\, \mu_i\, u_{(d+\frac{1}{2}),i,k} \qquad (4.1.42)$$

$$K_{(d+\frac{1}{2}),k} = \sum_i w_i\, \mu_i^2\, u_{(d+\frac{1}{2}),i,k} \qquad (4.1.43)$$

and

$$f_{(d+\frac{1}{2}),k}, \text{ the variable Eddington factor} = [K_{(d+\frac{1}{2}),k}/J_{(d+\frac{1}{2}),k}] \qquad (4.1.44)$$

where $d = (1, 2, \ldots, D)$ and w_i is the weight factor for appropriate quadrature.

In short, the scheme we have followed to determine the variable Eddington factor f_ν at all depths is to develop an iteration scheme for solving transfer equation starting with the source function taken as Planck function and calculate the angular moments and f_ν at all depths and frequencies. With f_ν, J_ν known at the end of first operation we can obtain a new source function S_ν which differs from B_ν. With this updated source function we may start the next iteration process until the convergence is secured. Eddington factor iteration converges fairly rapidly and the solution of transfer problem is substantially economical as compared to the outright use of Feautrier's scheme.

The Rybicki Method

Rybicki (1971) took note of the fact that in many problems of astrophysical interest, the scattering term is independent of frequency. In these cases, complete redistribution of photon takes place and the radiation field can be described effectively in terms of $\bar{J} = \int J_\nu \phi_\nu d\nu$ and one can avoid knowledge of frequency dependent informations at each depth. The method proposed by

4.1 Impact Parameter Method in Plane Geometry for Stationary Media

Rybicki for the solution of transfer problems contributes to the saving of computer time.

Rybicki dealt with vectors describing the depth dependence of the radiation field at specific frequency rather than specifying the frequency variation at a given depth. He defines the following two vectors $\mathbf{U}_k$ and $\bar{\mathbf{J}}$.

$$\mathbf{U}_k = (u_{3/2,k}, u_{5/2,k}, u_{7/2,k}\ldots, u_{(D+\frac{1}{2}),k}). \tag{4.1.45}$$

k is the specific angle-frequency point and

$$\bar{\mathbf{J}} = (\bar{J}_{3/2}, \bar{J}_{5/2}, \ldots, \bar{J}_{(D+\frac{1}{2})}). \tag{4.1.46}$$

Then the finite difference equation for the transfer equation (4.1.14) and the boundary conditions (4.1.23) and (4.1.24) yield for each angle-frequency point k, the system of equations of the form

$$\mathbb{T}_k\, \mathbf{U}_k + \mathbb{U}_k\, \bar{\mathbf{J}}_k = \mathbf{M}_k, \quad k = (1, 2, \ldots, K). \tag{4.1.47}$$

Here $\mathbb{T}_k$ is a $(D \times D)$ tridiagonal matrix which represent the differential operator at frequency k, $\mathbb{U}_k$ is a diagonal matrix containing the depth variation of scattering coefficient $\alpha_{(d+\frac{1}{2}),k}$, the source of function $S_{(d+\frac{1}{2}),k}$ is given by

$$S_{(d+\frac{1}{2}),k} = \alpha_{(d+\frac{1}{2}),k}\, \bar{J} + \beta_{(d+\frac{1}{2}),k}. \tag{4.1.48}$$

$\mathbf{M}_k$ is a vector containing the depth variation of thermal term in the source function $\beta_{(d+\frac{1}{2}),k}$ at k. We have one equation for each k-point. Further, we have D equations defining

$$\bar{J}_{(d+\frac{1}{2})} = \sum_{k=1}^{K} w_{(d+\frac{1}{2}),k}\, u_{(d+\frac{1}{2}),k} \tag{4.1.49}$$

with $d = (1, 2, \ldots, D)$. These are equivalent to the matrix equation

$$\bar{\mathbf{J}} = \sum_{k=1}^{K} \mathbb{V}_k\, \mathbf{U}_k. \tag{4.1.50}$$

For any particular angle-frequency point k, the tridiagonal system (4.1.47) gives us

$$\mathbf{U}_k = \mathbb{T}_k^{-1}\, \mathbf{M}_k - (\mathbb{T}_k^{-1}\mathbb{U}_k)\bar{\mathbf{J}}. \tag{4.1.51}$$

Now substituting (4.1.51) in (4.1.50), we have

$$\left[\mathbb{I} + \sum_{k=1}^{K} (\mathbb{T}_k^{-1}\mathbb{U}_k)\mathbb{V}_k\right] \bar{\mathbf{J}} = \sum_{k=1}^{K} \mathbb{V}_k \, \mathbb{T}_k^{-1} \, \mathbf{M}_k.$$

This can be written in the form

$$\mathbb{C} \, \bar{\mathbf{J}} = \mathbf{D}. \qquad (4.1.52)$$

In (4.1.52), $\mathbb{C}$ is a full $D \times D$ matrix given by

$$\mathbb{C} = \mathbb{I} + \sum_{k=1}^{K} (\mathbb{T}_k^{-1}\mathbb{U}_k) \, \mathbb{V}_k \qquad (4.1.53)$$

and

$$\mathbf{D} = \sum_{k=1}^{K} \mathbb{V}_k \, (\mathbb{T}_k^{-1}\mathbf{M}_k). \qquad (4.1.54)$$

The system of equations (4.1.52) is solved for $\bar{\mathbf{J}}$ and through it we obtain S_k, the source function at k for all depths. From (4.1.48), we have

$$\mathbf{S}_k = \alpha_k \, \bar{\mathbf{J}} + \beta_k \qquad (4.1.55)$$

and from (4.1.51), we have $\mathbf{U}_k$.

The computing time for Rybicki's scheme is more favourable than that of Feautrier [cf. Mihalas and Mihalas (1984) p. 378]. However, Rybicki's method is available only when the scattering is coherent, where as Feautrier's scheme is applicable in non-coherent scattering systems as well.

4.2 Impact Parameter Method in Spherical Geometry for Stationary Media

In the preceding article we have outlined Feautrier's method for solving radiative transfer problems in slab medium and Rybicki's modification of the method.

It will be shown here that by judicious transformation of reference frame, the radiative transfer equations in spherical geometry written along the rays and their attendant boundary conditions can be put in a form when the second order differential equations of transfer exhibit formal similarity to those in slab geometry. This immediately suggests that the methods developed in the earlier article can be applied to transfer problems in spherical geometry with appropriate modifications.

The direct application of Feautrier's method for solving transfer problems even in stationary spherically symmetric media may be time consuming as the order of matrices arising out of the discretisation process is rather large. But for the case of axially symmetric, isotropically scattering, spherical envelope the moment method via variable Eddington factor is simpler to use and it reduces the number of dependent variables involved. We shall review below this method proposed by Hummer and Rybicki (1971).

The radiative transfer equation in stationary spherically symmetric media in steady state can be written as [cf. (1.3.8) with absorption and emission coefficients angle independent].

$$\mu \frac{\partial I(r,\mu,\nu)}{\partial r} + \frac{1-\mu^2}{r} \frac{\partial I(r,\mu,\nu)}{\partial \mu} = -\chi_\nu(r)\left[I(r,\mu,\nu) - S(r,\nu)\right]. \tag{4.2.1}$$

The first two angular moments of the equation (4.2.1) can be written as

$$\frac{\partial}{\partial \tau_\nu}(r^2 H_\nu) = r^2(J_\nu - S_\nu) \tag{4.2.2}$$

and

$$\frac{\partial K_\nu}{\partial \tau_\nu} - \frac{1}{\chi_\nu r}(3K_\nu - J_\nu) = H_\nu, \tag{4.2.3}$$

with

$$d\tau_\nu = -\chi_\nu(r)\,dr. \tag{4.2.4}$$

χ_ν, H_ν, K_ν and J_ν are all functions of r and ν.

The variable Eddington factor $f_\nu(r)$ is defined by

$$f_\nu(r) = K_\nu(r)/J_\nu(r). \tag{4.2.5}$$

Using (4.2.5), (4.2.3) can be written as

$$[\partial(f_\nu J_\nu)/\partial \tau_\nu] - (3f_\nu - 1)J_\nu/(\chi_\nu r) = H_\nu. \tag{4.2.6}$$

Attempt is now made to put the combined equations (4.2.2) and (4.2.6) in a second order differential equation form which has formal similarly with (4.1.34) for plane parallel medium. This can then be shaped into a two point boundary value problem suitable for solution by Feautrier or Rybicki technique.

To do this a sphericity factor q_ν is introduced defined by

$$\ln q_\nu = \int_{r_c}^{r} [(3f_\nu - 1)/(r' f_\nu)] \, dr', \tag{4.2.7}$$

where r_c is the radius of the core star.

The equation (4.2.6) reads as

$$\frac{\partial (f_\nu q_\nu J_\nu)}{\partial \tau_\nu} = q_\nu H_\nu. \tag{4.2.8}$$

Substituting this in (4.2.2), we have the combined moment equation as

$$\frac{1}{q_\nu} \frac{\partial}{\partial \tau_\nu} \left[\frac{r^2}{q_\nu} \frac{\partial}{\partial \tau_\nu} (f_\nu q_\nu J_\nu) \right] = \frac{r^2}{q_\nu} (J_\nu - S_\nu). \tag{4.2.9}$$

In (4.2.9) changing the variable to X_ν where X_ν is given by

$$dX_\nu = \frac{q_\nu}{r^2} d\tau_\nu, \tag{4.2.10}$$

we have for (4.2.9)

$$\frac{\partial^2}{\partial X_\nu^2} [f_\nu q_\nu J_\nu] = \frac{r^4}{q} [J_\nu - S_\nu]. \tag{4.2.11}$$

The equation (4.2.11) is formally similar to (4.1.6).

The Boundary Conditions

The boundary condition at upper boundary $r = R$ can be obtained from (4.2.8) written at $r = R$ with the change of variable from τ_ν to X_ν. Thus the boundary condition at the upper surface $r = R$ can be expressed as

$$\left. \frac{\partial (f_\nu q_\nu J_\nu)}{\partial X_\nu} \right|_{r=R} = h_\nu \, R^2 J_\nu(R), \tag{4.2.12}$$

where

$$h_\nu = H_\nu(R)/J_\nu(R). \tag{4.2.13}$$

4.2 Impact Parameter Method in Spherical Geometry for Stationary Media

At the lower boundary, we invoke the diffusion approximation at the surface of the core star, $r = r_c$ and write

$$H_\nu(r_c) = \left(\frac{1}{3}\right)\left[\frac{1}{\chi_\nu(r)}\left|\frac{\partial B_\nu}{\partial r}\right|\right]_{r=r_c}. \tag{4.2.14}$$

In (4.2.14) the gradient is fixed to satisfy the condition that $H_\nu(r_c)$ integrated over all the frequencies equals the correct flux

$$H_c = (L_{\ell um}/16\pi^2 r_c^2), \tag{4.2.15}$$

where $L_{\ell um}$ is the luminosity.

Then the lower boundary condition reads

$$\frac{\partial(f_\nu q_\nu J_\nu)}{\partial X_\nu}\bigg|_{r=r_c} = r_c^2 H_c \left[\frac{1}{\chi_\nu}\left(\frac{\partial B_\nu}{\partial r}\right)\right/\int_0^\infty \frac{1}{\chi_\nu}\left(\frac{\partial B_\nu}{\partial r}\right)d\nu\bigg]_{r=r_c}. \tag{4.2.16}$$

The boundary condition at $r = r_c$ can differ depending on the physical nature of the core.

We are now in a position to use Feautrier's method to solve the two-point boundary value problem posed by (4.2.11), (4.2.12) and (4.2.16). For this we introduce a discrete radial mesh $\{r_d\}(d = 1, 2, \ldots D;$ with $r_1 = R$ and $r_D = r_c)$ and a frequency mesh $\{\nu_n\}(n = 1, 2, \ldots, N)$. We express, now, (4.2.11), (4.2.12) and (4.2.16) as finite-difference equations using splines or Hermite formulae. If and when the frequency integrals occur in the source function, it is replaced by appropriate quadrature formula. Then the combined difference equations corresponding to (4.2.11), (4.2.12) and (4.2.16) are written as a block tridiagonal system similar to that obtained in the preceding section and this system of equations is solved by Gaussian elimination scheme. In essence, we are using here the impact parameter method of Feautrier which is detailed in the preceding section for transfer problems in slab medium. In case the source function is frequency independent, one can use Rybicki type of method.

However, to carry out the above computation apriori knowledge of the variable Eddington factor $f_\nu(r) = K_\nu(r)/J_\nu(r)$ and the values of the angular moments at the boundaries of the atmosphere is essential. This is done by solving the transfer equation frequency by frequency along the rays tangent to a set of spherical shells using the given values of source functions. The angular information of the radiation field at each depth point is to be found out.

The transfer equation at each frequency has to be solved ray-by-ray along tangent to each discrete radial shell.

The (r, θ) or (r, μ) coordinates in axially symmetric spherical medium is changed to (s, p) coordinates along the rays [cf. Fig. 4.1]. s is measured positive from the axis of symmetry towards the direction of the observer and negative in the opposite direction. p is the impact parameter.

$$r = r(s, p) = (s^2 + p^2)^{1/2}. \tag{4.2.17}$$

Note is taken of the fact that the differential operator

$$\mu \frac{\partial}{\partial r} + \frac{1 - \mu^2}{r} \frac{\partial}{\partial \mu} = \frac{\partial}{\partial s}.$$

Writing the intensity of the rays travelling in the $\pm s$ as $I^\pm$, the transfer equation in spherical geometry along the rays is given by

$$\pm [\partial I^\pm(s, p, \nu)/\partial s] = \chi(s, p, \nu) [S(s, p, \nu) - I^\pm(s, p, \nu)]. \tag{4.2.18}$$

We now define $d\tau(s, p, \nu)$, $u(s, p, \nu)$ and $v(s, p, \nu)$ as

$$d\tau(s, p, \nu) = -\chi(s, p, \nu) ds, \tag{4.2.19}$$

$$u(s, p, \nu) = \frac{1}{2} [I^+(s, p, \nu) + I^-(s, p, \nu)], \tag{4.2.20a}$$

$$v(s, p, \nu) = \frac{1}{2} [I^+(s, p, \nu) - I^-(s, p, \nu)]. \tag{4.2.20b}$$

Now adding and subtracting the equations (4.2.18) and using (4.2.19) and (4.2.20), we have

$$[\partial v(s, p, \nu)/\partial \tau(s, p, \nu)] = u(s, p, \nu) - S(s, p, \nu) \tag{4.2.21}$$

and

$$[\partial u(s, p, \nu)/\partial \tau(s, p, \nu)] = v(s, p, \nu). \tag{4.2.22}$$

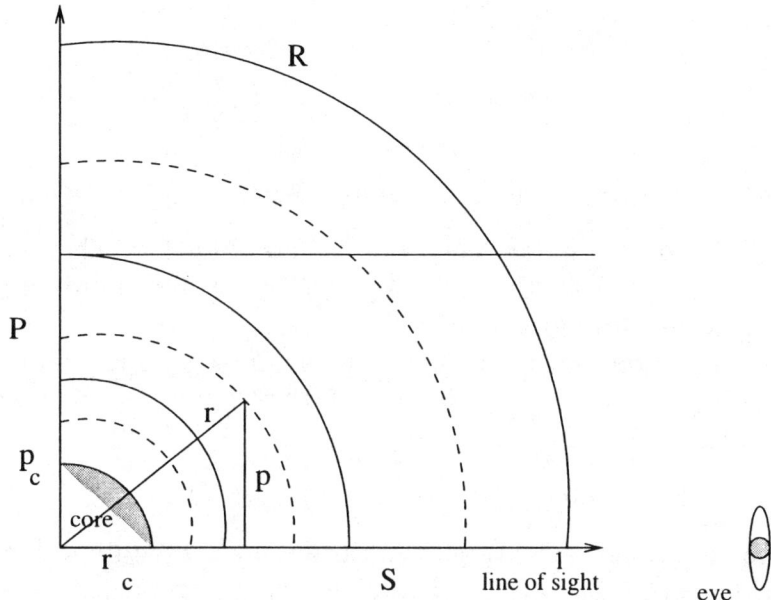

Figure 4.1. The p-s mesh used to solve the transfer equation in spherical geometry

Combining (4.2.21) and (4.2.22) we are led to the second order transfer equation

$$[\partial^2 u(s,p,\nu)/\partial \tau^2(s,p,\nu)] = u(s,p,\nu) - S(s,p,\nu). \quad (4.2.23)$$

The discretisation of s can be done through depth points $\{r_d\}(d = 1, 2, \ldots, D; r_D = r_c = s_c$ and $r_1 = R = s_{\max})$ and p through $\{p_i\}(i = 1, 2, \ldots, I; p_c = r_c = p_D, p_1 = p_R)$. $I = D + C$ where C stands for the number of rays which intersect the core star.

The Boundary Conditions

The boundary conditions under which (4.2.23) is to be solved are given below.

(i) The boundary condition at $r = R$, is $I^-(R,p,\nu) = 0$ which implies that $u(s_1,p,\nu) = v(s_1,p,\nu)$.

Then from (4.2.22)

$$\partial u(s,p_i,\nu)/\partial \tau(s,p_i,\nu)\Big|_{s_{\max}} = u(s_{\max},p_i,\nu) \qquad (4.2.24)$$

where $s_{\max} = (R^2 - p_i^2)^{1/2}$.

(ii) The inner boundary condition is determined by whether the ray is intercepted by the core $r_D = r_c$ or not and intersects the plane passing through $s = 0$.

For rays intersecting the core, the symmetry of the radiation field about the plane $s = 0$ suggest that $v(0,p_i,\nu) = 0$ which implies from (4.2.22)

$$\partial u(s,p_i,\nu)/\partial \tau(s,p_i,\nu)\,|_{s=0} = 0. \qquad (4.2.25)$$

The equations (4.2.23)–(4.2.25) are written as finite-difference equations and then combined to give a single tridiagonal system of equations. These are solved by Feautrier's algorithm detailed in the preceding section of this chapter. The source function is assumed to be known.

We obtain

$$u_{d,i,n} = u(s_d,p_i,\nu_n)$$

where $d = (1,2,\ldots,D);\ i = (1,2,\ldots,I;\ I = D+C)$ and $n = (1,2,\ldots,N)$.

When it is known, we can determine the angular moments, the mean intensity J and K-integral, using appropriate quadrature

$$J_{dn} = \sum_{i=1}^{i_d} w_{d,n}^{(0)}\, u_{d,i,n} \qquad (4.2.26)$$

$$K_{dn} = \sum_{i=1}^{i_d} w_{d,n}^{(2)}\, u_{d,i,n}, \qquad (4.2.27)$$

where w's are the appropriate quadrature weights.

To obtain $u_{d,i,n}$, it is represented by a piecewise polynomial and the weights determined by analytical integration of u in the $\{\mu_i\}$ mesh. We should note that a particular ray with an impact parameter p_i intersects a radial shell r_d at an angle $\cos^{-1}\mu_{d,i}$, where

4.2 Impact Parameter Method in Spherical Geometry for Stationary Media

$$\mu_{d,i} = (r_d^2 - p_i^2)^{1/2}/r_d = \frac{s_d}{r_d}. \qquad (4.2.28)$$

The variable Eddington factor $f_{d,n}$ at a particular depth and frequency is given by

$$f_{d,n} = K_{d,n}/J_{d,n}. \qquad (4.2.29)$$

We have already noted that to solve the transfer equation (4.2.11), we need prior knowledge of $f_{d,n}$. To overcome this difficulty, one starts an iteration scheme with an initial guess of Eddington factor $f_{d,n}$. With that assumed Eddington factor, the transfer equation (4.2.23) through (4.2.24) and (4.2.25) is solved to obtain $J_{d,n}$, $K_{d,n}$ from (4.2.26) and (4.2.27) and hence $f_{d,n}$ from (4.2.29). With this new $f_{d,n}$, the two point boundary value problem is solved again and the process is repeated until the convergence is secured. In most cases, the iteration scheme ends pretty rapidly.

An Alternative Method: Discrete Space Scheme

In the above scheme, to solve the radiative transfer equation, the source function S has to be given. We outline below the discrete space method [cf. Mihalas and Mihalas (1984) p. 383] as an alternative to Feautrier's scheme.

For simplicity, we take the participating medium to be spherically symmetric, grey and isotropically scattering.

The transfer equation (4.2.1) is rewritten as

$$3\mu \frac{\partial(r^2 I)}{\partial(r^3)} + \frac{1}{r}\frac{\partial}{\partial \mu}[(1-\mu^2)I] = -\chi[I - S]. \qquad (4.2.30)$$

We write (4.2.30) for $\pm\mu$ and take the sum and difference of these equations to get

$$3\mu \frac{\partial(r^2 v)}{\partial(r^3)} + \frac{1}{r}\frac{\partial}{\partial \mu}[(1-\mu^2)v] = -\chi(u - S) \qquad (4.2.31)$$

$$3\mu \frac{\partial(r^2 u)}{\partial(r^3)} + \frac{1}{r}\frac{\partial}{\partial \mu}[(1-\mu^2)u] = -\chi v, \qquad (4.2.32)$$

where u and v are defined in (4.1.2). The computational instability of μ-derivatives in (4.2.31) and (4.2.32) is avoided by integrating the equations over the cells $[\mu_{i-1/2}, \mu_{i+1/2}]$.

$$\mu_{i+1/2} - \mu_{i-1/2} = w_i, \quad i = 1, 2, \ldots, I \tag{4.2.33}$$

where w_i is the quadrature weight associated with μ_i. The number of quadrature points is I. Further, we take $\mu_{-1/2} = \mu_{I+1/2} = 0$.

Taking the above factors into account, (4.2.31) and (4.2.32) are written as [cf. Mihalas and Mihalas (1984) equations (83.100) and (83.104)]

$$3w_i\mu_i \left[\frac{\partial(r^2 v_i)}{\partial(r^3)}\right] + \frac{1}{r}\left\{(1 - \mu_{i+1/2}^2)v_{i+1/2} - (1 - \mu_{i-1}^2)v_{i-1/2}\right\}$$
$$= w_i\chi(S - u_i) \tag{4.2.34}$$

$$3w_i\mu_i^2 \left[\frac{\partial(r^2 u_i)}{\partial(r^3)}\right] + \frac{w_i(\mu_i^2 - 1)}{r}u_i$$
$$+ \frac{1}{r}\left\{\mu_{i+1/2}\left[1 - \mu_{i+1/2}^2\right]u_{i+1/2} - \mu_{i-1/2}\left[1 - \mu_{i-1/2}^2\right]\right\}u_{i-1/2}$$
$$= -\chi(w_i\mu_i v_i). \tag{4.2.35}$$

A linear spline approximation is used to replace $u_{i\pm 1/2}$ and $v_{i\pm 1/2}$ by their nodal values.

The resulting differential equations along with the boundary conditions are discretised to obtain a tridiagonal system (Mihalas and Mihalas (1984) p. 385). This discrete space scheme avoids the formal solution of transfer equation and involves a rather favourable use of computer time as compared to the impact parameter method.

4.3 Impact Parameter Method for Transfer Problems in Spherically Symmetric Moving Media

The solution of radiative transfer problems in lab frame for a radially expanding spherically symmetric medium was done by Kunasz and Hummer (1974). Their calculations are confined to

4.3 Impact Parameter Method for Spherically Symmetric Moving Media

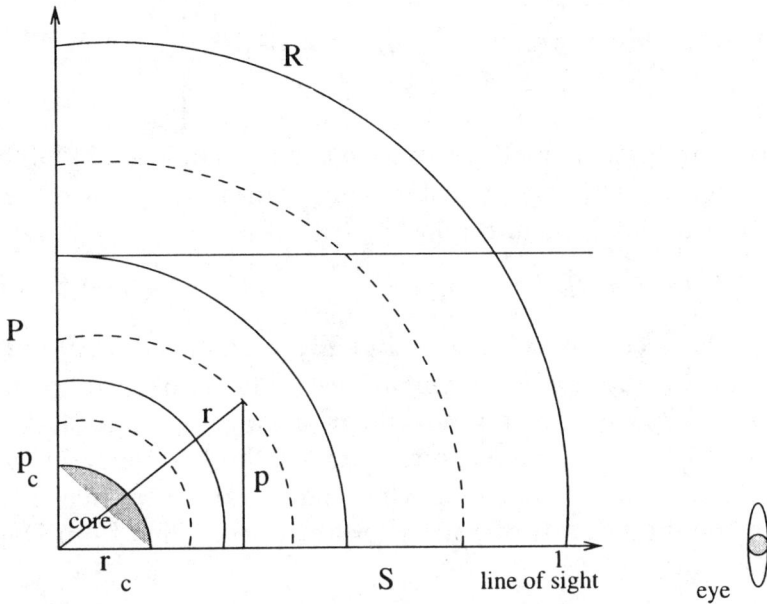

Figure 4.2. The p–s mesh used to solve the transfer equation in spherical geometry

the study of line formation in the interior layers of an atmosphere where the velocity of expansion is not too large. The radiation emanating from a central star permeating through a radially expanding envelope and received by an observer stationary in lab frame is studied. The reference frame is specified in (s, p, x) coordinates where x is the frequency in dimensionless unit given by $x = \frac{\nu - \nu_0}{\Delta \nu_D}$, with ν_0, the line centre frequency and $\Delta \nu_D$, the Doppler width. This reference frame was introduced in (3.1.16). We recall the Fig. 4.1.

p is the perpendicular distance of a particular ray from a ray passing through the centre towards the observer, the direction of the rays being parallel to each other, s is the distance along the ray considered positive towards the observer and negative away from him. In the (s, p, x) system, a ray-by-ray solution is obtained

and we have
$$\begin{aligned} s &= r\cos\theta = r\mu \\ p &= r\sin\theta = r(1-\mu^2)^{1/2} \\ r &= (s^2+p^2)^{1/2} \\ \mu &= s/(s^2+p^2)^{1/2}. \end{aligned} \quad (4.3.1)$$

We now write down the radiative transfer equation in lab frame for a radially expanding, spherically symmetric atmosphere. In Feautrier's (1964) form it can be written as

$$\pm[\partial I^\pm(s,p,x)/\partial s] = \eta(s,p,x) - \chi(s,p,x)I^\pm(s,p,x). \quad (4.3.2)$$

Let the medium be moving radially with speed $V(r)$ with respect to an external observer at rest. The photon frequency is Doppler shifted between the frame of the observer and that of the atoms constituting the matter. Let ν be the frequency in the lab frame and ν', the frequency in the atom's frame at which the photon in the direction $\mathbf{n}$ is emitted or absorbed. Then from Doppler effect

$$\nu' = \nu - \nu_0 \left(\frac{\mathbf{n}\cdot\mathbf{V}(r)}{c}\right). \quad (4.3.3)$$

Then in (4.3.2), the emissivity $\eta(s,p,x)$ and opacity $\chi(s,p,x)$ become angle dependent. It is customary to measure the frequency displacement from the line centre and express them in dimensionless units x given by

$$x = (\nu - \nu_o)/\Delta\nu_D, \quad (4.3.4)$$

with the Doppler width $\Delta\nu_D = (\nu_0 v_{th})/c$, where v_{th} is the thermal velocity. The velocity $V(r)$ is also written in dimensionless unit.

$$V = v(r)/v_{th}. \quad (4.3.5)$$

The relation between frequencies in lab frame and atom's frame is given by

$$x' = x - \mu V, \quad \text{with } \mu = \mathbf{n}\cdot\mathbf{V}. \quad (4.3.6)$$

Then the explicit form of (4.3.2) can be written as

$$\pm[\partial I^\pm(s,p,x)/\partial s] = \eta(s,p,x) - \chi(s,p,x)I^\pm(s,p,x), \quad (4.3.7)$$

where $\eta(s,p,x)$ and $\chi(s,p,x)$ are given by

$$\chi(s,p,x) = \chi_\ell(r)\phi(s,p,x) \quad (4.3.8)$$

4.3 Impact Parameter Method for Spherically Symmetric Moving Media

$$\eta(s, p, x) = \eta_\ell(r)\phi(s, p, x) \qquad (4.3.9)$$

with the profile function $\phi(s, p, x)$ defined by

$$\phi(s, p, x) = \phi[r(s, p); \ x - \mu(s, p)\, V(r)]. \qquad (4.3.10)$$

(r, μ) and (s, p) are connected by (4.3.1). It may be further noted $0 \leq \mu \leq 1$. $V(r)$ is positive in the direction of increasing r.

We now arrange to use Feautrier type of method for solving the problem. For this we define the mean intensity and flux like variable u and v respectively.

$$u(s, p, x) = \frac{1}{2}\left[I^+(s, p, x) + I^-(s, p, x)\right] \qquad (4.3.11)$$

and

$$v(s, p, x) = \frac{1}{2}\left[I^+(s, p, x) - I^-(s, p, x)\right]. \qquad (4.3.12)$$

The optical depth τ is given by

$$\tau(s, p, x) = \int_s^{s_{\max}} \chi(s', p, x)\, ds'. \qquad (4.3.13)$$

Equation (4.3.2) represents two first order differential equations.

Adding and subtracting these equations, we obtain two first order differential equations involving u and v. On eliminating v from them we obtain a second order differential equation in u as the equation of radiative transfer. It reads as

$$[\partial^2 u(s, p, x)/\partial \tau^2(s, p, x)] = u(s, p, x) - S(s, p, x) \qquad (4.3.14)$$

where the source function $S(s, p, x) = \eta(s, p, x)/\chi(s, p, x)$. In (4.3.13) we have assumed that $d\tau$ and χ are symmetric in μ and x.

The line source functions usually contain contributions from thermal and scattering terms. The scattering term is of non-local character and can be considerably affected by the motion of the medium. If complete redistribution of photons is assumed, the source function for two level atoms can be written as

$$S(s, p, x) = (1 - \varepsilon)\bar{J}(r) + \varepsilon B \qquad (4.3.15)$$

with

$$\bar{J}(r) = \int_{-x_{max}}^{x_{max}} dx \int_0^1 d\mu \phi[r, x - \mu V(r)] \, u[s(r,\mu), p(r,\mu), x]. \tag{4.3.16}$$

We note that in (4.3.16) ϕ is angle dependent and moreover in moving media the symmetry of the radiation intensity about the line centre cannot be taken for granted. Under these circumstances, in the study of radiative transfer in moving media, the validity of complete redistribution of photons is in doubt. Furthermore, the evaluation of the scattering integral is complicated by the interlocking of frequency and angle variations in $(x - \mu V)$. This creates difficulty in using frequency and angle quadrature in calculating the scattering integral. We can get rid of this draw back by expressing the radiative transfer equation in a coordinate frame comoving with the atom constituting the medium.

The line source function in a participating spherically symmetric moving atmosphere consisting of two level atoms, assuming complete redistribution of photons can be written as

$$\begin{aligned} S_\ell(s) &= (1-\varepsilon) \int_{-x_{max}}^{x_{max}} dx \int_0^1 u[s(r,\mu), p(r,\mu), x] \\ &\quad \times \phi(r, x - \mu V) \, d\mu + \varepsilon B(s) \end{aligned} \tag{4.3.17}$$

where ε is the line parameter and $B(s)$, the Planck function.

In (4.3.14), the transfer equation has been set as a second order differential equation in u, the solution of which is set in a medium confined between two spherical surfaces. So we are faced here with a two point boundary value problem, where the boundary conditions at the two bounding surfaces are to be spelt out. The following boundary conditions can be written down following the entire length of the ray $(-s_{max}, s_{max})$.

(a) For rays not intersecting the core, the incident intensities at the upper and lower boundaries $[\pm s_{max}]$ must be specified. This implies

$$[\partial u(s,p,x)/\partial \tau(s,p,x)]_{s=\pm s_{max}} = \pm u(s,p,x)|_{s=\pm s_{max}}. \tag{4.3.18}$$

(b) For rays intersecting the core star $(p \leq r_c)$.

4.3 Impact Parameter Method for Spherically Symmetric Moving Media

(i) $v(s_{\min}, p, x) = 0$, when the core is opaque. (4.3.19)

(ii) $[\partial u(s, p, x)/\partial \tau(s, p, x)]_{s=\pm s_{\min}} = \pm u(s, p, x)|_{s=\pm s_{\min}}$. (4.3.20)

To solve this problem, we use Feautrier type of method which has been explicitly demonstrated in sections (4.1) and (4.2). The discrete meshes $\{r_d\}$ and $\{p_i\}$ are introduced as in the static case. The frequency mesh $\{x_n\}$, $n = \pm 1, \pm 2, \ldots, \pm N$, however is extended over the whole profile with the symmetry property $x_{-n} = -x_n$ imposed. The finite-difference equations for transfer equations and boundary conditions are shaped into a matrix equation form suitable for application of Rybicki's modified method. $\bar{J}(r_d)$ is then determined. It should be noted that when we change the coordinates to (s, p, x) $u_{d,i,n} = u(s_d, p_i, x_n)$ is to be evaluated on a $\{s_{d,i}\}$ mesh along the complete length of the ray. When set in the Rybicki formalism [cf. (4.1.47)], the tridiagonal matrix $\mathbb{T}$ is square and matrix $\mathbb{U}$ is a Chevron matrix and rectangular. Then for each choice of (i, n), we obtain a solution of the type

$$\mathbb{U}_{i,n} = \mathbb{A}_{i,n}\,\bar{J} + \mathbb{B}_{i,n}. \quad (4.3.21).$$

Here $\bar{J}$ is discretised as

$$\bar{J}(r_d) = \sum_{x=-N}^{x=+N} w_n \sum_{i=1}^{I_d} a_{d,i}\phi[r_d;\, x_n - \mu(r_d, p_i)V(r_d)]\, u_{d,i,n},$$

(4.3.22)

where w_n and $a_{d,i}$ are quadrature weights. Now invoking the spherical symmetric nature of the radiation field, we may write

$$\left.\begin{array}{l} I^{\pm}(s, p, x) = I^{\pm}(-s, p, x) \\ u(s, p, -x) = u(-s, p, x) \\ v(s, p, -x) = -v(-s, p, x). \end{array}\right\} \quad (4.3.23)$$

Using (4.3.23) in (4.3.22), we may eliminate some of the variables and write $\bar{J}$ in terms of positive x and negative s and write

$$\begin{aligned}\bar{J}(r_d) &= \sum_{n=1}^{N} w_n \sum_{i=1}^{I_d} a_{d,i}\, \{\phi[r_d;\, x_n - \mu_{d,i}V_d]\, u_{d,i,n} \\ &\quad + \phi[r_d;\, x_n + \mu_{d,i}V_d]\, u_{d',i,n}\}\end{aligned} \quad (4.3.24)$$

with $d' = Di + 1 - d$.

As in (4.1.50), (4.3.24) can be written as a matrix equation

$$\bar{\mathbf{J}} = \sum_{k=1}^{K} \mathbb{V}_k \, \mathbf{U}_k \qquad (4.3.25)$$

where $k = i, n$ and $K = IN$. The $\mathbb{V}$ matrices in this case are Chevron matrices of rectangular type.

For obtaining the final system for $\bar{J}$, $u_{d,i,n}$ is calculated from (4.3.21) for all values of angle frequency points i and n and these system of equations is solved to obtain $\bar{J}$.

The method described here is stable. However, except for velocities of a few multiples of thermal velocities, the computing time needed is formidable. For large velocities, the comoving frame methods are found to be superior to those using lab frame. Kunasz and Hummer (1974) demonstrated the above scheme for some specialised situations such as when the opacity follows a power law and the velocity field, linear laws.

References

Feautrier, P. (1964): Compt. Rend., **258**, 3189.
Hummer, D.G., Rybicki, G. (1971): MNRAS, **152**, 1.
Kunasz, P.B., Hummer, D.G. (1974): MNRAS, **166**, 57.
Mihalas, D. (1978): Stellar Atmospheres. W. Freeman and Co., San Francisco.
Mihalas, D., Mihalas, B.W. (1984): Foundations of Radiation Hydrodynamics. Oxford University Press, New York.
Rybicki, G. (1971): JQSRT, **11**, 589.

Chapter 5. Numerical Methods for Transfer Equations in Lab Frame

We consider here some of the basic numerical methods for solving radiative transfer equations in spherically symmetric moving media when the transfer equations are expressed with respect to fixed observer's frame. To solve the relevant transfer equations, one has the option of treating the radiative transfer equation in its differential form or its integral form. In the differential equation approach, the radiation-matter interaction is considered local while in the integral equation approach the radiation-matter interaction at different positions are inter-locked through the integral equation. In chapter 4 we dealt with the differential form of the radiative transfer equation. In the present chapter we shall be concerned with the methods using integral form of the transfer equation.

In what follows we shall give an overview of the methods used in studying the radiative transfer problems in moving media, viz. direct quadrature method with linear interpolation, escape probability method and the core saturation method.

5.1 Direct Quadrature Method with Linear Interpolation

In this method an integral operator connecting the specific intensity I and the mean intensity $\bar{J}$ is used. The mean intensity is then related to the source function S by an auxiliary equation. A judicious choice of a quadrature representation is made to solve the transfer problem in the moving media. This method was first proposed by Avrett and Loeser (1984).

From equation (1.2.3), the radiative transfer equation along a specified ray is given by

$$\frac{dI(r,\mathbf{s},\nu)}{\chi(r,\mathbf{s})ds} = -I(r,\mathbf{s},\nu) + S(r,\nu). \qquad (5.1.1)$$

Assuming complete redistribution, let $\phi(\nu)$ be the line profile function for both the absorption and emission of radiation. We further assume that the absorption function is of the form

$$\chi(r,\nu) = \chi(r)\phi(\nu).$$

Introducing the monochromatic optical depth τ_ν by the relation

$$d\tau_\nu = -\chi(r)\phi(\nu)ds,$$

the transfer equation (5.1.1) can be written in the form

$$\frac{dI_\nu}{d\tau_\nu} = I_\nu - S(\tau_\nu). \qquad (5.1.2)$$

The formal solution of the above equation is

$$I_\nu = \int S(\tau'_\nu)\exp(-\tau'_\nu)\,d\tau'_\nu + C. \qquad (5.1.3)$$

We shall now describe the impact parameter method for an expanding spherical atmosphere. This is the method proposed by Avrett and Loeser (Kalkofen (1984) p. 341).

Consider a spherically symmetric atmosphere bounded by $r = r_o$ and $r = R$. The atmosphere is assumed to expand radially. We form a spherical grid with concentric circles of radii r_i, $i = 0, 1, 2, ..., N$. The optical depth τ_ν increases in the direction opposite to the radial distance with $\tau_\nu = 0$ at the outer boundary $r = R$. We classify the rays through an arbitrary point P as either a "shell-ray" or a "disc-ray". A ray is called a shell-ray if it does not intersect the inner boundary. A disc-ray crosses the inner boundary. As the standard boundary condition for spherically expanding atmospheres is the absence of incident radiation at the outer boundary, we set $C = 0$ for the shell-rays. For the disc-ray the value of C depends on the limits and the boundary condition at the inner boundary.

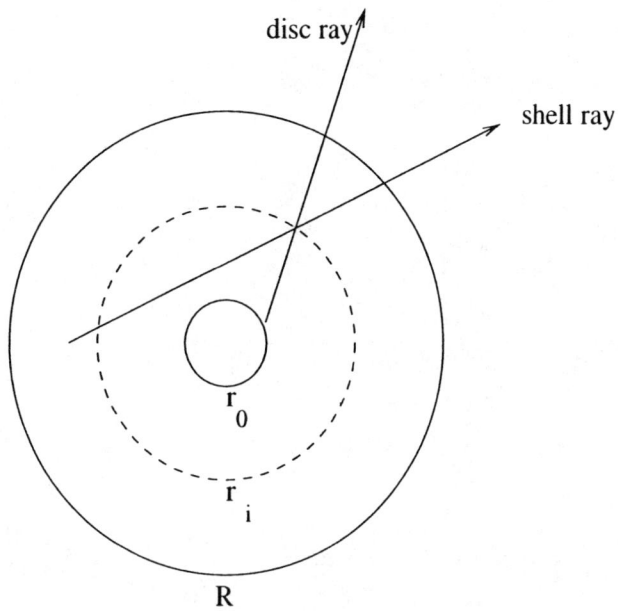

Figure 5.1. Direction of shell-ray and disc-ray in an envelope surrounding a core star

At the point r_i, let us denote by I_i^+ and I_i^- the outward and inward intensity respectively. If μ_{im} is the cosine of the angle between the outward ray (characterised by m) at r_i and the outward normal $\hat{n}$ at r_i, then

$$\mu_{im} = sign\left[1 - \left(\frac{r_m}{r_i}\right)^2\right]^{1/2}, \qquad (5.1.4)$$

where r_m is the radius of the circle having the given ray with directions **s** as the tangent. $m < N$ for a shell ray and $m = \alpha r_N\, 0 \leq \alpha \leq 1$. Also

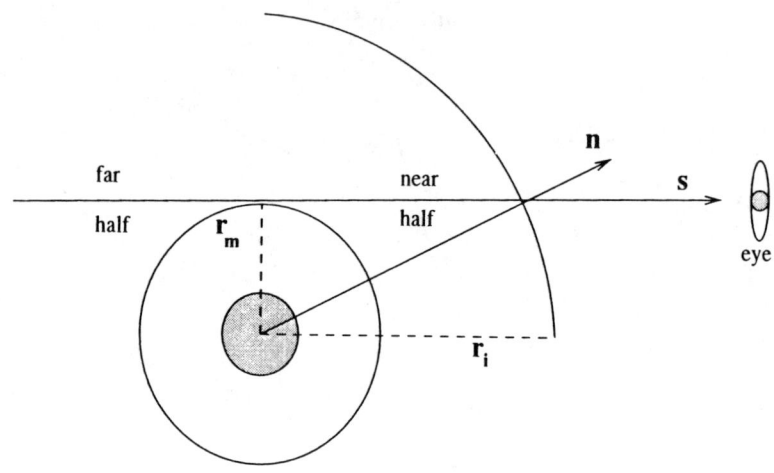

Figure 5.2. Direction of a tangent ray in an envelope surrounding a core star

$$\text{sign} = \begin{cases} +1, & \text{for disc-ray} \\ +1, & \text{for the near-half of shell-ray} \\ -1, & \text{for the far-half of shell-ray.} \end{cases}$$

In the case of a stationary atmosphere, the line profile function $\phi(\nu)$ is a known function, usually taken to be either a Voigt profile or a Doppler profile. In the case of a Doppler profile

$$\phi(\nu) = \frac{1}{\sqrt{\pi}\Delta\nu_D}\exp(-u^2), \qquad (5.1.5)$$

where $u = \dfrac{\nu - \nu_0}{\Delta\nu_D}$, is the frequency displacement from the line centre ν_o in units of Doppler half-width $\Delta\nu_D$. Using wave-length displacement instead, we have

$$u = \frac{\lambda - \lambda_0}{\Delta\lambda_D} \qquad (5.1.6)$$

where
$$\Delta\lambda_D = \frac{\lambda_0}{c}\left(\frac{2kT}{M} + V_b^2\right)^{1/2}. \tag{5.1.7}$$

Here T is the temperature; M, the atomic mass; and V_b, the assumed turbulent velocity. Now that we are considering an expanding atmosphere moving with radial velocity $V(r)$ towards the observer, the line profile function depends on u in the same way as before but u is given by the new Doppler shifted relation

$$u = \frac{(\lambda - \lambda_0) + \left(\frac{\lambda_0}{c}\right)V(r)}{\Delta\lambda_D}. \tag{5.1.8}$$

If we now consider a point r_i moving with radial velocity V_i and if μ_{im} is the cosine of the angle between the ray and the normal at r_i, then we have

$$u_{ikm} = \frac{(\lambda_k - \lambda_0) + \left(\frac{\lambda_0}{c}\right)\mu_{im}V_i}{\Delta\lambda_D}, \tag{5.1.9}$$

as $\mu_{im}V_i$ is the velocity component along the given ray. Here the suffix k denotes the frequency grid points. Unlike the stationary atmosphere, in the case of an expanding atmosphere it is essential to distinguish between the outward and inward rays as the velocity component will be of opposite signs. We also realise that the sign of μ must be changed between the outward and inward rays in precisely the same manner. As the effect of the velocity component is to shift the lines either to the red or blue end, we can disregard the distinction between outward and inward rays and regard the inward ray term $(\lambda_k - \lambda_0)$ as $-(\lambda_0 - \lambda_k)$. This coupled with a change in sign of μ_{im} has no effect on $\phi(\mu_{ikm})$ as μ_{ikm} occurs in $\phi(\mu_{ikm})$ as a squared term. With this understanding we can now ignore the difference between the outward and inward rays but must allow both positive and negative values for $(\lambda_k - \lambda_0)$. If we now consider the profile at r_i as seen in the co-moving frame at r_ℓ, i.e. as seen by an observer located at r_ℓ and moving radially with velocity V_ℓ, the new profile ϕ will depend on

$$u_{ikm}^\ell = \frac{(\lambda_k - \lambda_0) + \left(\frac{\lambda_0}{c}\right)[\mu_{im}V_i - \mu_{\ell m}V_\ell]}{(\Delta\lambda_D)_i}. \tag{5.1.10}$$

This will help us to determine the optical distance from point r_ℓ to any other point r_i along the ray, as seen in the comoving frame at r_ℓ. In the subsequent discussion we will suppress the index ℓ which refers to the frame of observation and denote the expression on the r.h.s of equation (5.1.10) as just u_{ikm}. There will be no confusion in the final computation with the expression in (5.1.9) as we will usually take $V_\ell = 0$.

Let us now define a dimensionless radial velocity $\bar{V}_i$ at r_i in terms of the velocity V_i at r_i

$$\bar{V}_i = \frac{V_i}{c}\left(\frac{\lambda_0}{\Delta\lambda_D}\right).$$

Letting $u_k = \left(\dfrac{\lambda_k - \lambda_0}{\Delta\lambda_D}\right)$, we have from (5.1.6)

$$u_{ikm} = u_k + \left(\mu_{im}\bar{V}_i - \mu_{\ell m}\bar{V}_\ell\right). \tag{5.1.11}$$

We can now compute the optical depth τ_{ikm} of the shell of radius r_i as viewed from r_ℓ at wavelength λ_k in the direction μ_{im}. We have

$$\tau_{ikm} = -\int \chi(r)\phi(u_{ikm})\left(\frac{dr}{\mu_{im}}\right).$$

The mean intensity $\bar{J}_i$ at r_i is given by [cf. (3.2.4)]

$$\bar{J}_i = \frac{1}{2}\int_{-1}^{1} d\mu \int_{-\infty}^{\infty} \phi(u_{ikm}) I(r_i, \mu, \lambda)\, d\lambda.$$

First we evaluate the wavelength integral by a quadrature and obtain

$$\int_{-\infty}^{\infty} \phi(u_{ikm}) I(r_i, \mu_{im}, \lambda)\, d\lambda = \sum_k A_k \phi_k(u_{ikm}) I_{ikm}$$
$$= \tilde{I}_{im}(\text{say}). \tag{5.1.12a}$$

This quadrature is selected such that the normalization condition of the profile

$$\sum_k A_k \phi_k(u_{ikm}) = 1 \tag{5.1.12b}$$

is satisfied.

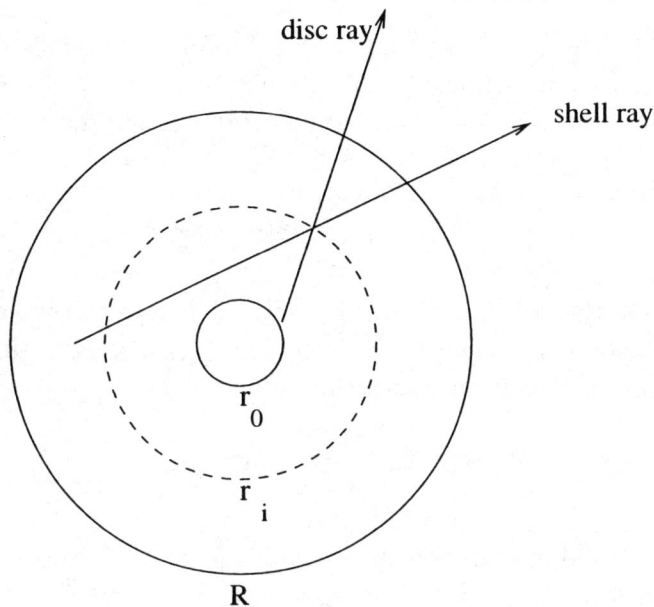

Figure 5.3. Direction of an arbitrary intensity ray in a spherical envelope

The mean intensity integral is then computed assuming that it varies linearly with μ between the values I_{ikm} at μ_{im}, $0 < \mu_{im} \leq 1$, with $\mu_{i1} = 0, \mu_{iM} = 1$, where $m = 1, 2, \ldots, M$. Thus

$$\bar{J}_i = \sum_{m=1}^{M} c_{im} \tilde{I}_{im}, \qquad (5.1.13)$$

where the coefficients c_{im} are given by

$$c_{im} = \frac{1}{2} \begin{cases} (\mu_{i2} - \mu_{i1}), & m = 1 \\ (\mu_{i(m+1)} - \mu_{i(m-1)}), & m \neq 1 \\ (\mu_{iM} - \mu_{i(M-1)}), & m = M. \end{cases}$$

We observe that the angle quadrature coefficients satisfy the normalization condition

$$\sum_{m}^{M} c_{im} = 1.$$

Auxiliary Equation for S and $\bar{J}$

In this analysis we shall use the standard model of a two level atom with complete redistribution in frequency. From equation (1.4.16) we have the auxiliary equation connecting the source function S and the mean intensity $\bar{J}$ given by

$$S_i = \frac{\bar{J}_i + \epsilon_i B_i}{1 + \epsilon_i}, \qquad (5.1.14)$$

where B_i is the Planck function and ϵ_i is given by equation (1.4.17). Also from the formal solution (5.1.3), assuming S is isotropic, we obtain with a suitable quadrature

$$I_{ikm} = \sum_{j=1}^{N} \bar{W}_{ijkm} S_j.$$

It is found expedient to introduce the unit matrix $\mathbb{I}$ and write the above equation as

$$I_{ikm} - S_i = \sum_{j=1}^{N} W_{ijkm} S_j, \qquad (5.1.15)$$

where $W_{ijkm} = \bar{W}_{ijkm} - \delta_{ij}$. From (5.1.11), (5.1.12a), (5.1.12b), (5.1.13) and (5.1.15) we have

$$\bar{J}_i = \sum_m^M c_{im} \left(\sum_k A_k \phi_k(u_{ikm}) \left\{ S_i + \sum_j^N W_{ijkm} S_j \right\} \right)$$

$$\bar{J}_i = \sum_k A_k \left(\sum_j \widetilde{W}_{ijk} S_j \right) + S_i,$$

where $\widetilde{W}_{ijk} = \sum_m c_{im} W_{ijkm} \phi_k(u_{ikm})$. Now let us set $\sum A_k \widetilde{W}_{ijk} = \widetilde{W}_{ij}$ and write the above equation as

$$\bar{J}_i = \sum_j \widetilde{W}_{ij} S_j + S_i. \qquad (5.1.16)$$

Now from (5.1.14) and (5.1.16) we obtain the required equation for the source function S_i as

$$(1 + \epsilon_i)S_i = S_i + \sum_j \widetilde{W}_{ij}S_j + \epsilon_i B_i,$$

i.e.
$$S_i - \frac{1}{\epsilon_i} \sum_j \widetilde{W}_{ij} S_j = B_i. \qquad (5.1.17)$$

The above quadrature scheme will be complete once we know how to obtain the quadrature weighting coefficients W_{ijkm} in equation (5.1.15). From the formal solution (5.1.3) we have for the shell ray $C = 0$,

$$\sum_{j=1}^{N} W_{ijkm} S_j = \frac{1}{2} \int_0^{\tau_{\mathcal{N}km}} S(t) exp(-|t - \tau_{ikm}|)\, dt - S_i \qquad (5.1.18)$$

where τ_{ikm} is the optical depth and $\mathcal{N} = 2N$ for the present shell-ray. Assuming again that $S(t)$ varies linearly between each pair of grid points τ_{jkm} and $\tau_{(j+1)km}$ and defining

$$\Delta_{jkm} = \tau_{(j+1),km} - \tau_{j,km}, \qquad j = 1, 2, \ldots, N \qquad (5.1.19)$$

and $\Delta_{0km} = \Delta_{\mathcal{N}km} = 0$, we have

$$W_{ijkm} = \frac{1}{2} \begin{cases} e^{-(\tau_i - \tau_{j+1})} q(\Delta_j) + e^{-(\tau_i - \tau_j)} m(\Delta_{j-1}), & j < 1 \\ m(\Delta_{i-1}) + m(\Delta_i) - 2, & j = 1 \\ e^{-(\tau_{j-1} - \tau_i)} q(\Delta_{j-1}) + e^{-(\tau_j - \tau_i)} m(\Delta_j), & j > 1 \end{cases}$$
$$(5.1.20)$$

where

$$m(\Delta) = 1 - \frac{1 - e^{-\Delta}}{\Delta}$$

$$q(\Delta) = 1 - \frac{1 - e^{-\Delta}}{\Delta} - e^{-\Delta}.$$

Here we have suppressed the indices km on the right hand side. For the disc-ray in semi-infinite medium, $\mathcal{N} = N$ and we add the contributions $W'_{iN-1} = -K_i$ and $W'_{iN} = (1 + \Delta_{N-1})K_i$ to the last two columns of j obtained above with

$$K_i = \frac{1}{2\Delta_{N-1}} e^{-(\tau_N - \tau_i)}. \qquad (5.1.21)$$

Further refinement of the weights have been proposed by Avrett and Loeser to account for the difference spacing of the path length with the angle μ_{im} [cf. Kalkofen (1984) p. 378.].

5.2 Escape Probability Method

In chapter 3 we have introduced the concept of escape probability P_e. In this section we will show how this concept is used to obtain an approximate solution of the transfer equation. From (3.2.4) the mean integrated intensity $\bar{J}$ is given by

$$\bar{J}_{(r)} = \frac{1}{4\pi} \int d\Omega \int_0^\infty d\nu \phi \left[\nu - \frac{\nu_0}{c}\hat{s} \cdot \bar{V}(r)\right] I(r, s, \nu),$$

the integration being over all frequencies and solid angles. Using the formal solution (5.1.3), we have

$$\bar{J}(r) = \frac{1}{4\pi} \int d\Omega \int_0^\infty \phi \left[\nu - \frac{\nu_0}{c}\hat{s} \cdot \bar{V}(r)\right] d\nu \int S(\tau') \exp(-\tau'_\nu) d\tau'_\nu. \tag{5.2.1}$$

It is useful to express $\bar{J}$ as a volume integral over the source function. That is done by observing that $s = |\mathbf{r} - \mathbf{r}'|$, and $d^3\mathbf{r}' = s^2 ds\, d\Omega$. We have

$$\bar{J} = \int K(\mathbf{r}, \mathbf{r}') S(\mathbf{r}') \chi(\mathbf{r}') d^3\mathbf{r}', \tag{5.2.2}$$

where the kernel function $K(\mathbf{r}, \mathbf{r}')$ of this operator is given by

$$K(\mathbf{r}, \mathbf{r}') = \frac{1}{4\pi|\mathbf{r} - \mathbf{r}'|^2} \int_0^\infty \phi(\hat{s}, \nu, \mathbf{r}) \phi(\hat{s}, \nu, \mathbf{r}') \exp(-\tau(\mathbf{r}, \mathbf{r}', \nu)) d\nu,$$

where the monochromatic optical depth of the ray from $\mathbf{r}$ to $\mathbf{r}'$ is

$$\tau(\mathbf{r}, \mathbf{r}', \nu) = \int_0^{|\mathbf{r}-\mathbf{r}'|} \chi(\bar{\mathbf{r}}) \phi\left(\nu' - \frac{\nu_0}{c}\hat{s} \cdot \bar{V}(r')\right) ds$$

with $\bar{\mathbf{r}} = \mathbf{r} - s\hat{s}$ and $\hat{s} = \dfrac{\mathbf{r} - \mathbf{r}'}{|\mathbf{r} - \mathbf{r}'|}$. It is possible now to relate the escape probability function P_e defined in (3.1.2) to the kernel function $K(\mathbf{r}, \mathbf{r}')$. The quantity $K(\mathbf{r}, \mathbf{r}')\chi(\mathbf{r}')d\mathbf{r}'^3$ is the probability that a photon emitted at point $\mathbf{r}$ will be absorbed by the volume element $d^3\mathbf{r}'$ about $\mathbf{r}'$. Since the escape probability at $\mathbf{r}$ is equal to the probability that the photon emitted at $\mathbf{r}$ will **not** be absorbed anywhere in the medium, we have

$$P_e(\mathbf{r}) = 1 - \int K(\mathbf{r}, \mathbf{r}') \chi(\mathbf{r}') d^3\mathbf{r}'. \tag{5.2.3}$$

5.2 Escape Probability Method

The first order escape probability method is based on the following approximation. Suppose that the scale of variation of the source function S is very much larger than the scale of variation of the kernel function K, then it is reasonable to assume S is nearly a constant in equation (5.2.1) for the mean intensity and pull out S outside the integral. Thus from (5.2.1) and (5.2.2) we get in this approximation

$$\bar{J} = [1 - P_e(\mathbf{r})] S(r). \tag{5.2.4}$$

This is called the first order escape probability approximation relating $\bar{J}$ and S. For static atmospheres, the first order escape probability approximation is reasonably good at large depths but becomes bad near the surface. Consider the standard two-level atom model for which we have

$$S = (1 - \epsilon)\bar{J} + \epsilon B. \tag{5.2.5}$$

Now (5.2.3) and (5.2.4) give

$$S = \frac{\epsilon B}{\epsilon + (1 + \epsilon)P_e(\mathbf{r})}. \tag{5.2.6}$$

In the semi-infinite plane parallel static atmospheres the source function given in (5.2.6) approaches B at large depths as it should. But at the surface where the escape probability is $1/2$, the above approximation (5.2.5) gives

$$S(0) = \frac{2\epsilon B}{1 + \epsilon}.$$

This is very different from the exact value $S(0) = B\sqrt{\epsilon}$, especially when ϵ is very small. Thus there was an attempt to improve this approximation so that it was reasonably valid even near the surface. This attempt, at present has been successful only for the semi-infinite plane parallel medium when the profile function ϕ is depth independent. For this case, one can obtain [cf. Rybicki (1984) p. 56] the following exact equation

$$\int_\sigma^\infty \frac{\partial \bar{J}}{\partial \tau} S(\tau) d\tau = \int_\sigma^\infty d\tau S(\tau) \frac{\partial}{\partial \tau} \int_0^\sigma d\tau' K_1(|\tau - \tau'|)S(\tau') + \frac{1}{2}S_\infty^2, \tag{5.2.7}$$

where $K_1(\tau) = \dfrac{1}{2}\int_0^\infty d\nu \phi(\nu) E_1(|\tau|)\phi(\nu)$. Assuming again that the scale of variation of S is much larger than that of the kernel function K_1, we can obtain the desired approximation by pulling both $S(\tau)$ and $S(\tau')$ outside the integral. Using the fact that $\dfrac{\partial K_1}{\partial \tau}(|\tau - \tau'|)$ goes to zero fast for large separation of τ' and τ, it is seen that [cf. Rybicki (1984) p. 57.]

$$\frac{\partial \bar{J}}{\partial \tau} = \frac{\partial S}{\partial \tau} - \frac{\partial P_e}{\partial \tau} S - 2P_e \frac{\partial S}{\partial \tau}.$$

Integrating this and using the boundary condition $J(\infty) = S(\infty)$, we have

$$\bar{J}(\tau) = (1 - P_e)S(\tau) + \int_\tau^\infty dt P_e \frac{\partial S}{\partial t}. \qquad (5.2.8)$$

This is called by Hummer and Rybicki (1982) as the second order escape probability approximation. This approximation does in fact give a good approximation at both boundaries.

In the above, we have a physical analysis of the escape probability method and its connection with Sobolev's theory for solving radiative transfer problems in media moving with high speed. A modified form of core saturation method of Rybicki (1972) is used iteratively to obtain the exact solution of the problem. An outline of the core saturation method follows.

5.3 Core Saturation Method

This method was proposed by Rybicki (1972) and was first demonstrated to solve radiative transfer problems in static slab media. There is a close relationship between this method and the escape probability approach and the probabilistic equation of Scharmer (1981). The essential feature of Rybicki's method is to develop a numerical structure for solving the radiative transfer equation in statistical equilibrium using the relationship between the source function and the mean intensity. He assumed that for large monochromatic depth, the source function could be taken to be equal to the mean intensity at the line core. The relationship between the source function and the mean intensity at the wings was determined, as usual, from the radiative transfer equation. Rybicki's core saturation scheme was used by Stenholm (1977). Stenholm (1980) used it for comoving frame in moving media and Stenholm and Stenflo (1977) used it in cylindrical geometry. Scharmer (1981) utilised it to develop the probabilistic equations. Rybicki distinguished between "saturated core region" and "wing" in terms of frequencies. He noted that when the monochromatic intensity for all frequencies ν is equal to the source function S, i.e. $I(\nu) = S$, the mean intensity $\bar{J}$ saturates to S, i.e. $\bar{J}(\nu) = S$ for all frequencies. The basis of Rybicki's (1972) core saturation method is to replace the idea of complete saturation by partial saturation and assume that $I(\nu) = S$ for only a particular set of frequencies depending on the ray under consideration at the specified location of the point. The set of frequencies defines the core region or more explicitly "saturated core region". The frequencies outside this set are said to belong to the "wing".

The saturated core region is defined in terms of the monochromatic optical depth being greater than some assigned value γ.

$$I(\nu) \simeq S \quad \text{for} \quad \tau_{\nu b} > \gamma;$$

where $\tau_{\nu b}$ is the monochromatic optical depth from the boundary. The value of γ determines the extent of the line that is taken to be saturated. When $\gamma \to \infty$, all intensities belong to the wing. For general computation γ is assumed to be of the order of unity.

In a depth-frequency grid, at any depth one looks at the whole set of frequencies. At each depth a critical frequency x_c is defined such that

$$\tau \phi(x_c) = \gamma,$$

where ϕ is the profile function and x_c is the dimensionless critical frequency.

$$|x| < x_c \implies \text{core frequency}$$
$$|x| > x_c \implies \text{wing frequency.}$$

Kalkofen and Ulmschneider (1984) extended the core saturation method for solving transfer problems in moving media. We describe below the essential features of this method.

We recall from equation (3.2.4), that the mean intensity is given by

$$\bar{J}(\tau) = \frac{1}{4\pi} \int d\Omega \int_0^\infty d\nu\, \phi\left[\nu - \frac{\nu_0}{c}\hat{s} \cdot \mathbf{V}(r)\right] I(\tau, \mathbf{s}, \nu),$$

where the integration is over all frequencies and solid angles. For a spherically symmetric atmosphere we can rewrite this as

$$\bar{J}(\tau) = \frac{1}{2} \int_{-1}^{+1} d\mu \int_0^\infty d\nu\, \phi\left[\nu - \frac{\nu_0}{c}\hat{s} \cdot \mathbf{V}(r)\right] I(\tau, \mu, \nu).$$

The basis of the core saturation method is to split the frequency integral into two parts called the core and wing regimes and approximate the intensity in terms of the source function in these two regions. In the case of an expanding spherical atmosphere, the frequencies are affected, depending on the component of the velocity in the direction of observation. Thus rewriting the above expression for the line core and line wings we have

$$\bar{J} = \frac{1}{2} \int_{\nu,\mu \in core} d\nu \int d\mu \phi I + \frac{1}{2} \int_{\nu,\mu \in wings} d\nu \int d\mu \phi I. \quad (5.3.1)$$

Let us denote the contribution by the wing frequencies for a unit intensity by $\bar{J}_0(\tau)$. Then by the normalization condition, the core-contribution for a unit intensity is $1 - \bar{J}_0(\tau)$,

i.e. $$\frac{1}{2} \int_{\nu,\mu \in core} d\nu \int d\mu \phi = 1 - \bar{J}_0(\tau).$$

In the core saturation approximation, the assumption is that the value of the intensity in the line core is equal to the local value of the line source function,

i.e. $$I(\tau, \mathbf{s}, \nu) = S(\tau), \qquad \nu \in \text{core}.$$

With this assumption, equation (5.3.1) can be written as

$$\begin{aligned}\bar{J}(\tau) &= S(\tau)\left[1 - \bar{J}_0(\tau)\right] + \frac{1}{2} \int_{\nu,\mu \in wings} d\nu \int d\mu \phi \\ &\quad \times \left[\nu - \frac{\nu_0}{c}\hat{s} \cdot \mathbf{V}(r)\right] I(\tau, \mu, \nu). \end{aligned} \quad (5.3.2)$$

For the standard two level atom with complete redistribution we have, as before [cf. 5.1.10]

$$S = \frac{\bar{J} + \epsilon B}{1 + \epsilon}.$$

Using (5.3.2) in this we have

$$S(\tau) = \frac{\frac{1}{2}\int_{\nu,\mu \in wings} d\nu \int d\mu \phi I + \epsilon B}{\bar{J}_0(\tau) + \epsilon}. \quad (5.3.3)$$

Using the formal solution (5.1.3), the specific intensity $I(\tau, \mu, \nu)$ for the wing frequencies can be expressed in terms of the source function in a layer whose thickness is of the order of a photon mean free path. Thus equation (5.3.3) forms an integral equation for the source function.

We will now indicate how one can determine the core-wing separation. Given a spatial grid τ_i, a photon at depth τ_i is regarded to be in the core or in a wing of the line if the monochromatic optical distance to the next grid point in the upstream direction for the photon under discussion is larger or smaller, respectively, than γ, an optical parameter of order unity. Using the standard convention in which the inward normal $\mu = -1$, points along the positive τ axis, the monochromatic optical distance, δ, can be calculated easily. For the outward flowing photons between the spatial grid points τ_i and τ_{i+1}, we have

$$\delta = \frac{1}{\mu_m} \bar{\phi} \times (\tau_{i+1} - \tau_i), \quad \mu_m > 0$$

where $\bar{\phi}$ is the average profile function in the interval $(i, i+1)$ given by

$$\bar{\phi} = \frac{1}{2}[\phi(\tau_i, \nu_k, \mu_m) + \phi(\tau_{i+1}, \nu_k, \mu_m)]$$

for the photons in the ray (ν_k, μ_m). Similarly for the inward flowing photons

$$\delta = \frac{1}{\mu_m} \tilde{\phi} \times (\tau_i - \tau_{i-1}), \quad \mu_m > 0$$

where $\tilde{\phi} = 1/2[\phi(\tau_i, \nu_k, \mu_m) + \phi(\tau_{i-1}, \nu_k, \mu_m)]$. In the above ν_k is the discretized frequency displacement, $\nu - \nu_0$, from the rest frequency ν_0 of the line.

If $\delta > \gamma$, then the photons are considered to be in the line core, otherwise they belong to the wings. This criteria used at each grid point and for given direction, is essential for the expanding atmospheres. Some numerical work can be reduced by noting that if photons at (τ_i, ν_k, μ_m) belong to the core then so will be the photons at $(\tau_i, -\nu_k, -\mu_m)$.

We will now discuss the approximate method suggested by Kalkofen and Ulmschneider (1984) to solve the transfer equation (5.3.3) in the wings. Generalizing the well known Eddington-Barbier relations, they took the intensity at τ_i to be equal to the source function at unit optical distance from τ_i, i.e.

$$I(\tau_i, \nu_k, \mu_m) = S\left[\tau_i + \frac{\mu_m}{\phi(\tau_i, \nu_k, \mu_m)}\right]. \tag{5.3.4}$$

Since the argument in the source function do not, in general, coincide with any spatial grid point, it is necessary to interpolate. They suggested the use of the monochromatic optical distance δ defined earlier between grid points for this interpolation. Suppose δ_i^j denotes the optical distance along the ray (ν_k, μ_m) and $(-\nu_k, -\mu_m)$ between the spatial grid points τ_j and τ_i, then define

$$I^+(\tau_i, \nu_k, \mu_m) = \bar{\omega} S_j + (1 - \bar{\omega}) S_{j+1}$$

with

$$\bar{\omega} = (\delta_i^{j+1} - 1)/\delta_i^{j+1}$$

and

$$I^-(\tau_i, -\nu_k, -\mu_m) = \tilde{\omega} S_\ell + (1 - \tilde{\omega}) S_{\ell-1}$$

with

$$\tilde{\omega} = (\delta_{\ell-1}^i - 1)\delta_{\ell-1}^\ell.$$

The intervals $(j, j+1)$; $(\ell, \ell-1)$ are to be chosen such that they contain the source point $\tau_i \pm \dfrac{\mu_m}{\phi(\tau_i, \pm\nu_k, \pm\mu_m)}$ which are at unit optical distance along the ray from τ_i.

Once the intensity is expressed in terms of the source function, equation (5.3.3) has the form

$$S(\tau) = X(\tau, \tau')S(\tau') + Y(\tau), \tag{5.3.5}$$

with the angle and frequency integrated matrix operator X given by

$$X(\tau,\tau') = \frac{1}{\bar{J}_0(\tau) + \epsilon} \sum_{\nu,\mu \in wings} W_{\mu\nu} \phi \left[\nu - \frac{\nu_0}{c}\hat{s} \cdot \bar{V}(r)\right] exp(-\tau)$$

where $W_{\mu\nu}$ are the quadrature weights for the integration. This can be solved iteratively. In the next section we will detail the differential operator perturbation method which shows an efficient way to solve numerically the integral equation.

In the standard core saturation method the separation of line photons into core and wing photons is defined in terms of the optical depth of the given point. This tackles the escape of radiation from the surface of the medium but ignores transfer within the atmosphere. To take care of this aspect, the present separation looks at the monochromatic optical distance to the next grid point in the upstream direction.

Refrences

Avrett, E.H., Loeser, R. (1984) Methods in Radiative Transfer, ed. Kalkofen, W. Cambridge University Press, Cambridge, p 341.
Cannon, C.J. (1973): JQRST, **13**, 627.
Cannon, C.J. (1973): Ap. J., **185**, 621.
Hamann, W.R., (1987): Numerical Radiative Transfer, ed Kalkofen, W. Cambridge University Press, Cambridge, p. 23.
Hummer, D. G. Rybicki, G.B. (1982): Ap. J., **263**, 925.
Kalkofen, W., Ulmschneider, P. (1984): Methods in Radiative Transfer, ed Kalkofen, W. Cambridge University Press, Cambridge, p. 131.
Kalkofen, W. (1987): Numerical Radiative Transfer, Cambridge University Press, Cambridge.
Rybicki, G. B., (1972): Line Formation in the presence of Mag. Fields ed. Athay, R.G., Honse, L. L., Newkirk, G. Boulder High Altitude Observatory, p. 145.
Rybicki, G. B., (1984): Methods in Radiative Transfer, ed Kalkofen, W. Cambridge University Press, Cambridge, p. 21.
Scharmer, R. (1981): Ap. J., **249**, 720.
Scharmer, R. (1984): Methods in Radiative Transfer, ed Kalkofen, W. Cambridge University Press, Cambridge. p. 173.
Stenholm, L.G. (1977): Astron. Astrophys., **54**, 572.
Stenholm, L.G. (1980): Astron. Astrophys. Suppl., **42**, 23.
Stenholm, L.G., Stenflo, J.O. (1977): Astron. Astrophys. **58**, 273.

PART

III

Chapter 6. Impact Parameter Method for Transfer Equations in Comoving Frame (CMF)

In part II, we have described the methods of solving the radiative transfer equations in lab frame for a spherically symmetric stationary and moving media. We have noted that Doppler effect due to the motion of the stellar envelope with respect to the observer makes opacity and emissivity angle-dependent. This interlocking of angle and frequency in the scattering term of source function presents serious complications in using suitable quadrature formulae for angle and frequency. However, when the transfer equation is written in a comoving frame, the opacity and emissivity are isotropic and hence angle independent in such frames and the static redistribution functions of photons can be used even in the cases of partial redistribution. Moreover in comoving frame, band width for scattering calculations can be taken only to cover the line profile and it does not depend on fluid velocity. Angular quadrature can be taken in this frame to be similar to that in the static case.

In chapter 2, we have deduced the radiative transfer equations in comoving and mixed frames. We transformed the radiation field and material variables in transfer equations from lab frame to comoving or mixed frame by way of Lorentz transformation. Strictly speaking Lorentz transformation is applicable when the relative velocities between the frames under consideration are uniform. Within the scope of special relativity, the transfer equations expressed in two uniformly moving frames are covariant. However, the conditions prevailing in the moving stellar envelope seldom conforms to this limitation. The velocity of expansion and/or rotation of the material of the atmosphere are usually functions of radial distance from the centre of the star and may also be of time. To accommodate these conditions, account has to be taken of the change of Lorentz transformation in the medium from point to point. We can only take "local Lorentz transformation" between variables expressed in lab frame and those at successive elementary points of the medium with assigned velocities. In the transfer equation in comoving frames we must also take note of the presence of terms containing frequency derivatives of intensity. In solving them we have to deal with partial integro-differential equations with initial-plus-boundary values.

In the present chapter, we shall consider the impact parameter method of solution of radiative transfer problems in a spherically symmetric moving envelope when the transfer equations are written in comoving frame. Only the effect of Doppler shift will be considered and those of aberration and advection will be ignored. We shall review the works of Mihalas, Kunasz and Hummer (1975, 1976) and Mihalas *et al.* in the subsequent years.

6.1 Feautrier's Impact Parameter Method and Solution of Transfer Equations in Spherically Symmetric Comoving Frame

In section (4.3), we have stated the use of Feautrier method in studying the line formation problem in optically thick, spherically symmetric stellar envelope in radial motion with respect to the centre. The coordinate frame used in that case is of an observer at rest with respect to the star centre. In what follows, we shall use a modified form of Feautrier's method to solve the transfer equation in comoving frame [cf. Mihalas, Kunasz and Hummer (1975)].

The presence of velocity field in the medium gives rise to Doppler shift of photons and other angle dependent physical effects like aberration and advection each one of $0(V/c)$. However, the study of relative importance of these physical effects suggests that in the case of line profile calculation, Doppler effect plays the major role [cf. Mihalas (1978) p. 492]. The frequency shift in a line due to aberration and advection is $\frac{\Delta \nu}{\nu} \simeq V/c$, while the Doppler shift $\frac{\Delta \nu}{\Delta \nu_D} = (V/v_{th})$. This implies that in a variation of line profile with frequency, Doppler effect has more significant contribution than the effects of aberration and advection. In the present section, we shall ignore the effects of aberration and advection and retain only the contribution from Doppler effect.

The Transfer Equation in Comoving Frame and its Solution

Here we are interested in the study of transfer equation at r for radiation of frequency ν flowing in a direction $\cos^{-1}\mu$ with the radius vector of a spherically symmetric atmosphere surrounding a core star. It is recorded by an observer stationed at an element of medium at r, which is moving with velocity $V(r)$. The radiative transfer equation in the comoving frame can be written as [cf. equation (2.1.14)].

$$\mu_0[\partial I^0(r,\mu_0,\nu_0)/\partial r] + \frac{1-\mu_0^2}{r}[\partial I^0/\partial \mu_0]$$
$$-(\nu_0 V(r))/(cr)[(1-\mu_0^2)$$
$$+\mu_0^2(d\ln V/\partial \ln r)](\partial I^0(r,\mu_0,\nu_0)/\partial \nu_0)$$
$$= \eta^0(r,\nu_0) - \chi^0(r,\nu_0)I^0(r,\mu_0,\nu_0)$$
$$= -\chi^0(r_0,\nu_0)[I^0(r,\mu_0,\nu_0) - S^0(r,\nu_0)]. \quad (6.1.1)$$

In (6.1.1), the symbols have their usual meanings. The quantities with prefixes and suffixes zero indicate that they are being measured in comoving frame. η^0 and χ^0 are total emissivity and opacity respectively. The explicit forms of η^0 and χ^0 and hence the source function $S^0(r,\nu_0) = \eta^0/\chi^0$ are given below

$$\chi^0(r,\nu_0) = \chi_L(r)\,\phi^0(r,\nu_0) + \chi_c^0(r) + \sigma_e^0 n_e(r) \quad (6.1.2)$$

and

$$\eta^0(r,\nu_0) = \eta_L^0(r)\,\phi^0(r,\nu_0) + \eta_c^0(r) + \sigma_e^0 n_e(r) J_c^0(r). \quad (6.1.3)$$

The subscripts L and c refer to the line and continuum. n_e^0 and σ_e^0 are electron density and electron scattering cross-sections respectively; the subscript e signifying electron. Doppler width of the electron is assumed to be large and electron scattering is taken mainly to be due to continuum radiation.

In equations (6.1.1)–(6.1.3) all the quantities are expressed in comoving frame. Henceforward we shall drop the sub- and superscripts zero for labelling the quantities in comoving frame. This is not expected to give rise to any serious confusion.

The line opacity and emissivity are given by [cf. Mihalas, Kunasz and Hummer (1975) p. 273],

$$\chi_L(r) = \sigma_{\ell u}[n_\ell(r) - g_\ell n_u(r)/g_u] \qquad (6.1.4)$$

and

$$\eta_L(r) = \frac{2h\nu^3}{c^2} \sigma_{\ell u} g_\ell n_u(r)/g_u. \qquad (6.1.5)$$

In the above, the subscripts ℓ and u refer to the lower and upper states. $n_\ell(r)$ and $n_u(r)$ are the electron populations at r and g_ℓ and g_u are the statistical weights at the respective levels. $\sigma_{\ell u}$ is the scattering cross-section.

The opacity and emissivity for continuum are given by

$$\chi_c(r) = \sum_b \sigma_{bc}(\nu)[n_b(r) - n_b^*(r)e^{-h\nu/kT}] \qquad (6.1.6)$$

$$\eta_c(r) = \sum_b \frac{2h\nu^3}{c^2} e^{-h\nu/kT} \sigma_{bc}(\nu) n_b^*(r). \qquad (6.1.7)$$

Here the subscript b denotes the bound level of electrons and asterisked n's denote non-LTE populations. The summation extends over all bound states which are ionised by radiation of frequency ν.

The source function

$$S(r, \nu) = \frac{S_L(r)\phi(r, \nu) + \Psi}{\phi(r, \nu) + \bar{\omega}} \qquad (6.1.8)$$

where the line source function

$$S_L(r) = \eta_L(r)/\chi_L(r). \qquad (6.1.9)$$

Now taking note of the condition for statistical equilibrium, the source function for two level atoms with complete redistribution can be written as [cf. 3.1.37]

$$S(r, \nu) = (1 - \varepsilon)\bar{J}(r) + \varepsilon B(r), \qquad (6.1.10)$$

$B(r)$ is the Planck function at the line centre for electron temperature at r.

In (6.1.8)
$$\Psi = [\eta_c(r) + n_e(r)\sigma_e \bar{J}_c]/\chi_L(r) \tag{6.1.11}$$
$$\bar{\omega} = [\chi_c(r) + n_e(r)\sigma_e]/\chi_L(r). \tag{6.1.12}$$

Now going back to the equation of transfer (6.1.1) in comoving frame, we transform the spherical polar coordinates (r, μ) to (s, p) system introduced earlier. s is the distance along the ray measured positive towards the external observer, p the impact parameter [cf. Fig. 4.1], where
$$r = r(s, p) = (s^2 + p^2)^{1/2} \tag{6.1.13}$$
$$\mu = s/(s^2 + p^2)^{1/2}. \tag{6.1.14}$$

Then the transfer equation for radiation flow along a ray of frequency ν and impact parameter p can be written as
$$\pm[\partial I^{\pm}(s, p, \nu)/\partial s] - \bar{\gamma}(s, p)[\partial I^{\pm}(s, p, \nu)/\partial \nu]$$
$$= \eta[r(s, p), \nu] - \chi[r(s, p), \nu]I^{\pm}(s, p, \nu), \tag{6.1.15}$$

where we have written
$$\mu \frac{\partial}{\partial r} + \frac{1 - \mu^2}{r}\frac{\partial}{\partial \mu} = \frac{\partial}{\partial s} \tag{6.1.16}$$

and
$$\bar{\gamma}(s, p) \equiv \frac{\nu V(r)}{cr}\left[1 - \mu^2 + \mu^2\left(\frac{d \ln V}{d \ln r}\right)\right]. \tag{6.1.17}$$

We now use the Feautrier scheme of introducing the mean intensity and flux like variables u and v defined by
$$u(s, p, \nu) \equiv \frac{1}{2}[I^+(s, p, \nu) + I^-(s, p, \nu)] \tag{6.1.18}$$

and
$$v(s, p, \nu) \equiv \frac{1}{2}[I^+(s, p, \nu) - I^-(s, p, \nu)]. \tag{6.1.19}$$

In comoving frame, we can average I^+ and I^- at a given value of ν.

Now write
$$d\tau(s, p, \nu) = -\chi(s, p, \nu)ds \tag{6.1.20}$$

and
$$\gamma(s,p,\nu) = \bar{\gamma}(s,p)/\chi(s,p,\nu). \tag{6.1.21}$$

Substituting (6.1.18)–(6.1.21) in (6.1.15) and adding and subtracting the resultant equations, we have

$$[\partial u(s,p,\nu)/\partial \tau(s,p,\nu)] + \gamma(s,p,\nu)[\partial v(s,p,\nu)/\partial \nu] = v(s,p,\nu) \tag{6.1.22}$$

and

$$[\partial v(s,p,\nu)/\partial \tau(s,p,\nu)] + \gamma(s,p,\nu)[\partial u(s,p,\nu)/\partial \nu] = u(s,p,\nu) - S(s,p,\nu) \tag{6.1.23}$$

where from (6.1.10)

$$S(s,p,\nu) = S(r,\nu) = (1-\varepsilon)\bar{J}(r) + \varepsilon B(r) \tag{6.1.24}$$

and

$$\bar{J}(r) = \int_{\nu_{\min}}^{\nu_{\max}} d\nu\, \phi(\nu) \int_0^1 d\mu\, u[s(r,\mu), p(r,\mu), \nu] \tag{6.1.25}$$

$\nu_{\max}$ and $\nu_{\min}$ limits the complete line profile as observed in co-moving frame.

At this point, it is interesting to compare (6.1.24) and (4.3.16) for the nature of the profile function ϕ. In the comoving frame, the symmetry of the radiation field can be assumed while this cannot be done in the lab frame formulation.

The Boundary and Initial Conditions

(i) At the outer surface $r = R$, $I^- = 0$ (no incident radiation from outside). This implies that $u(R,\nu) = v(r,\nu)$. Hence from (6.1.22)

$$[\partial u(s,p,\nu)/\partial \tau(s,p,\nu)]_{s_{\max}} + \gamma(s,p,\nu)[\partial u(s,p,\nu)/\partial \nu]_{s_{\max}} = u(s_{\max},p,\nu). \tag{6.1.26}$$

(ii) At $s = 0$, which is the plane of symmetry $v(0,p,\nu) = 0$. Then from (6.1.22)

$$[\partial u(s,p,\nu)/\partial \tau(s,p,\nu)]_{s=0} = 0. \tag{6.1.27}$$

For the rays intersecting the core, one of the following things may happen.

(a) When the core is opaque as in the case of stellar surface, v is specified at $r = r_c$.

(b) When the core is transparent, by symmetry v can be taken to be zero at $s = 0$.

(iii) The initial condition for frequency:

For a star with an expanding atmosphere $[V > 0, \frac{dV}{dr} > 0]$, any photon to be intercepted at the $\nu_{\max}$ side of the line profile in comoving frame must originate at the continuum. The line photons emerging from any other point of the atmosphere will be red-shifted and escape interaction with the high frequency edge of the line profile.

The initial condition then can be written as

$$u(s, p, \nu_{\max}) = u_{\text{continuum}}. \qquad (6.1.28)$$

This can be obtained from (6.1.22) and (6.1.23) under the condition that terms like $\partial/\partial \nu$ are set equal to zero, thus giving the $u_{\text{continuum}}$ radiation, otherwise the term $(\partial u/\partial \nu)_{\nu_{\max}}$ must be specified. For example

$$(\partial u/\partial \nu)_{\nu_{\max}} = 0 \qquad (6.1.29)$$

implies that the slope of the continuum at $\nu_{\max}$ is zero.

The equations (6.1.22)–(6.1.29) give us the basic statement of the partial-initial-boundary value problem involved in the radiative transfer problems in expanding spherically symmetric stellar atmospheres.

The Method of Solution

The system of transfer equations and initial and boundary conditions are discretised using the grids

$$\begin{aligned} \{r_d\}, &\quad d = 1, 2, \ldots, D \\ \{p_i\}, &\quad i = 1, 2, \ldots, I \\ \{s_{d,i}\}, &\quad d, i = 1, 2, \ldots, DI \end{aligned}$$

and

$$\{\nu_n\}, \qquad n = 1, 2, \ldots, N.$$

We replace the integrals in (6.1.25) by quadrature sums and write $\bar{J}(r)$ as

$$\bar{J}(r_d) = \sum_{n=1}^{N} w_n \sum_{i=1}^{I_d} a_{d,i} \phi(r_d, \nu_n) - u[s(r_d, p_i), p_i, \nu_n]. \quad (6.1.30)$$

Then the equations (6.1.22) and (6.1.23) are written as finite difference equations. To do this, we assume that

(a) u is defined on the mesh points $s_d = s(r_d, p_i)$

$$u_{d,i,n} = u(s_d, p_i, \nu_n) \quad (6.1.31)$$

(b) v is defined on the mesh points $s_{d\pm 1/2} = \tfrac{1}{2}[s_d \pm s_{d\pm 1}]$

$$v_{(d\pm 1/2),i,n} = v(s_{d\pm 1/2}, p_i, \nu_n) \quad (6.1.32)$$

(c) $\chi_{(d\pm 1/2),i,n} = \tfrac{1}{2}[\chi_{d,i,n} + \chi_{(d\pm 1),i,n}]$ \hfill (6.1.33)

(d) $\Delta \tau_{d,i,n} = \tfrac{1}{2}[\Delta \tau_{(d+1/2),i,n} + \Delta \tau_{(d-1/2),i,n}]$ \hfill (6.1.34)

where

$$\Delta \tau_{(d+1/2),i,n} = \chi_{(d+1/2),i,n}[s_d - s_{(d+1)}] \quad (6.1.35)$$

$$\Delta \tau_{(d-1/2),i,n} = \chi_{(d-1/2),i,n}[s_{(d-1)} - s_d] \quad (6.1.36)$$

and

(e) $\delta_{d,i,(n-1/2)} = \gamma_{d,i,n}/(\nu_{n+1} - \nu_n).$ \hfill (6.1.37)

γ is defined in (6.1.21). Then the equations (6.1.22) and (6.1.23) are expressed as finite difference equations after discretisation of variables and use of relations (6.1.30) through (6.1.37). They appear as

$$[(u_{(d+1),i,n} - u_{d,i,n})/\Delta \tau_{(d+1/2),i,n}$$
$$= v_{(d+1/2),i,n} + \delta_{(d+1/2),i,(n-1/2)}$$
$$\times \left[v_{(d+1/2),i,n} - v_{(d+1/2),i,(n-1)}\right] \quad (6.1.38)$$

and
$$[(v_{(d+1/2),i,n} - v_{(d-1/2),i,n})/\Delta\tau_{d,i,n}]$$
$$= u_{d,i,n} - S_{d,i,n}$$
$$+\delta_{d,i,(n-1/2)}[u_{d,i,n} - u_{d,i(n-1)}]. \quad (6.1.39)$$

Now solving (6.1.38) for $v_{(d+1/2),i,n}$, we have

$$v_{(d+1/2),i,n} = \{[(u_{(d+1),i,n} - u_{d,i,n})/\Delta\tau_{(d+1/2),i,n}]$$
$$+ v_{(d+1/2),i,(n-1)}\delta_{(d+1/2),i,(n-1/2)}\}$$
$$/[1 + \delta_{(d+\frac{1}{2}),i,(n-1/2)}]. \quad (6.1.40)$$

The equation (6.1.40) is used to eliminate $v_{(d+1/2),i,n}$ from (6.1.39), and we obtain a second order system in $u_{d,i,n}$.

The system of equations read as

$$\frac{1}{\Delta\tau_{d,i,n}}\left[\frac{u_{(d+1),i,n}}{\Delta\tau_{(d+1/2),i,n}(1 + \delta_{(d+1/2),i,(n-1/2)})}\right]$$
$$+ u_{d,i,n}\left\{\frac{1}{\Delta\tau_{(d+1/2),i,n}(1 + \delta_{(d+1/2),i,(n-1/2)})}\right.$$
$$\left.+ \frac{1}{\Delta\tau_{(d-1/2),i,n}(1 + \delta_{(d-1/2),i,(n-1/2)})}\right\}$$
$$+ \frac{u_{(d-1),i,n}}{\Delta\tau_{(d-1/2),i,n}(1 + \delta_{(d-1/2),i,(n-1/2)})}$$
$$= (1 + \delta_{d,i,(n-1/2)})u_{d,i,n} - S_{d,i,n}$$
$$- \delta_{d,i,(n-1/2)}u_{d,i,(n-1)}$$
$$+ \frac{1}{\Delta\tau_{d,i,n}}\left[\frac{\delta_{(d-1/2),i,(n-1/2)}v_{(d-1/2),i,(n-1)}}{1 + \delta_{(d-1/2),i,(n-1/2)}}\right.$$
$$\left.- \frac{\delta_{(d+\frac{1}{2}),i,(n-1/2)}v_{(d+1/2),i,(n-1)}}{1 + \delta_{(d+1/2),i,(n-1/2)}}\right]. \quad (6.1.41)$$

The left-hand side of (6.1.41) contains u in three successive depth points at the identical frequency point. We get a set of second order equations for $u_{d,i,n}$ — second order in spatial derivatives and first order in frequency derivative. The boundary conditions are also written in the form of finite difference equations. The details of the procedure is given in Mihalas, Kunasz and Hummer (1975, p. 470–471).

Elimination Scheme

For the solution of the system of equations (6.1.41) and the appropriate boundary and initial conditions Gaussian elimination scheme of Feautrier can be used at depth point to depth point. However, in spherical medium the existence of peaking effect of radiation complicates matters. The number of rays in $\{p_i\} - \{\nu_n\}$ blocks to be used in that case may be formidable and Feautrier solution may prove costly. In these circumstances, Rybicki type of solution, under the condition of complete redistribution of photons [cf. section (4.4)] may be profitable. We shall outline this below.

The equation (6.1.40) represents the depth variation of v along a particular ray at a given frequency. This can be put in a vector form as
$$\mathbf{v}_{i,n} = \mathbb{G}_{i,n}\mathbf{u}_{i,n} + \mathbb{H}_{i,n}\mathbf{v}_{i,(n-1)}, \quad (6.1.42)$$
where
$$\mathbf{u}_{i,n} \equiv (u_{1,i,n}, u_{2,i,n}, \ldots, u_{D,i,n}) \quad (6.1.43)$$
and
$$\mathbf{v}_{i,n} \equiv (v_{1/2,i,n}, v_{3/2,i,n}, \ldots, v_{(D-1/2),i,n}). \quad (6.1.44)$$
In (6.1.42) $\mathbb{H}$ is a diagonal and $\mathbb{G}$ is a bidiagonal matrix.

Similarly, we express the equation (6.1.41) along with the finite difference equations for the boundary conditions in the form of a matrix equation system. We can write it as
$$\mathbb{T}_{i,n}\mathbf{u}_{i,n} + \mathbb{U}_{i,n}\mathbf{u}_{i,(n-1)} + \mathbb{V}_{i,n}\mathbf{v}_{i,(n-1)} + \mathbb{W}_{i,n}\bar{\mathbf{J}} = \mathbf{X}_{i,n}, \quad (6.1.45)$$

$\mathbf{X}_{i,n}$ is a vector. $\mathbb{T}_{i,n}$ is a tridiagonal matrix, $\mathbb{U}_{i,n}$ and $\mathbb{W}_{i,n}$ are diagonal matrices and $\mathbb{V}_{i,n}$ is a bidiagonal matrix.

The complete system of equations (6.1.45) are solved ray by ray, a particular ray being identified by a given impact parameter p_i. The frequency by frequency integration is carried out with n running from 1 to N.

The initial condition in frequency implies that the matrices $\mathbb{U}_{i,1}, \mathbb{V}_{i,1}$ and $\mathbb{H}_{i,1}$ are all exactly zero.

6.1 Feautrier's Impact Parameter Method

Then we can write from (6.1.42)–(6.1.45)

$$\mathbf{u}_{i,1} = \mathbf{A}_{i,1} - \mathbb{B}_{i,1}\bar{\mathbf{J}} \tag{6.1.46}$$

and

$$\mathbf{v}_{i,1} = \mathbf{C}_{i,1} - \mathbb{D}_{i,1}\bar{\mathbf{J}} \tag{6.1.47}$$

where

$$\mathbf{A}_{i,1} = \mathbf{X}_{i,1}\mathbb{T}_{i,1}^{-1} \tag{6.1.48}$$

$$\mathbb{B}_{i,1} = \mathbb{T}_{i,1}\mathbb{W}_{i,1}^{-1} \tag{6.1.49}$$

$$\mathbf{C}_{i,1} = \mathbb{G}_{i,1}\mathbf{A}_{i,1} \tag{6.1.50}$$

$$\mathbb{D}_{i,1} = \mathbb{G}_{i,1}\mathbb{B}_{i,1} \tag{6.1.51}$$

where $\mathbf{A}_{i,1}$ and $\mathbf{C}_{i,1}$ are vectors and $\mathbb{B}_{i,1}$ and $\mathbb{D}_{i,1}$ are matrices.

$u_{i,n}$ and $v_{i,n}$ for successive values of n can be obtained by similar substitutions.

Thus we get

$$\mathbf{u}_{i,n} = \mathbf{A}_{i,n} - \mathbb{B}_{i,n}\bar{\mathbf{J}} \tag{6.1.52}$$

$$\mathbf{v}_{i,n} = \mathbf{C}_{i,n} - \mathbb{D}_{i,n}\bar{\mathbf{J}} \tag{6.1.53}$$

with

$$\mathbf{A}_{i,n} = \mathbb{T}_{i,n}^{-1}[\mathbf{X}_{i,n} - \mathbb{U}_{i,n}\mathbf{A}_{i,(n-1)} - \mathbb{V}_{i,n}\mathbf{C}_{i,(n-1)}] \tag{6.1.54}$$

$$\mathbb{B}_{i,n} = \mathbb{T}_{i,n}^{-1}[\mathbb{W}_{i,n} - \mathbb{U}_{i,n}\mathbb{B}_{i,(n-1)} - \mathbb{V}_{i,n}\mathbb{D}_{i,(n-1)}] \tag{6.1.55}$$

$$\mathbf{C}_{i,n} = \mathbb{G}_{i,n}\mathbf{A}_{i,n} + \mathbb{H}_{i,n}\mathbf{C}_{i,(n-1)} \tag{6.1.56}$$

and

$$\mathbb{D}_{i,n} = \mathbb{G}_{i,n}\mathbb{B}_{i,n} + \mathbb{H}_{i,n}\mathbb{D}_{i,(n-1)}. \tag{6.1.57}$$

Along every ray p_i, $u_{i,n}$ is calculated for every frequency ν_n and this is substituted in (6.1.30) to secure the final system of equations for determining $\bar{J}$. This system can be written as [cf. Mihalas (1978) p. 508]

$$\left[\mathbb{I} + \sum_{i,n}\mathbb{F}_{i,n}\mathbb{B}_{i,n}\right]\bar{\mathbf{J}} = \sum_{i,n}\mathbb{F}_{i,n}\mathbf{A}_{i,n} \tag{6.1.58}$$

where $\mathbb{I}$ is the identity matrix.

The solution of (6.1.58) gives us $\bar{J}$ and hence $S(r,\nu)$. Then from (6.1.52) and (6.1.53), we get $u(s,p,\nu)$ and $v(s,p,\nu)$ and hence

$u(r,\mu,\nu)$ and $v(r,\mu,\nu)$. We can also calculate the mean intensity, $J^0(r,\nu)$, flux $F^0(r,\nu)$ and K^0-integral in the comoving frame.

In solving the above problem, Rybicki's scheme was preferred to Feautrier's in consideration of favourable computer time required in it. However, we have already noted that Rybicki's method is available only in the case of complete redistribution of photons whereas Feautrier's method can be used for partial redistribution as well. Feautrier's scheme can also be used for solving moment equations of intensity where Eddington factors are obtained by ray-by-ray formal solution with an initial estimate of source function.

The above method has been utilised to calculate the source functions and line profiles in a number of idealised problems [cf. Mihalas, Kunasz and Hummer (1975)]. It is, however, flexible and general enough to take care of more realistic problems. For the successful execution of the method, relevant radiation variables χ_L, χ_L/χ_c, $(n_e\sigma_e)/\chi_L$, line parameter ε, the Planck function B and the velocity V must be given. Further more, normalised line profile ϕ, the physical dimension of the stellar envelope, appropriate boundary and initial conditions are to be specified. With some judicious modifications the impact parameter method can be used to tackle a wide variety of problems.

6.2 The Role of Aberration and Advection in Radiative Transfer Problems in Spherically Symmetric Moving Media

The importance of aberration and advection terms in the formulation and solution of radiative transfer equations in spherically symmetric moving medium was examined by Mihalas, Kunasz and Hummer (1976). The transfer equation is written in the comoving frame retaining terms involving aberration and advection, the orders of which are (V/c). This form was also used by Castor (1972).

The radiative transfer equation in spherically symmetric moving medium in steady state in comoving frame keeping terms upto $0(V/c)$ is given by equation (2.5.10)

6.2 The Role of Aberration and Advection

$$\left[\mu + \frac{V(r)}{c}\right]\frac{\partial I(r,\mu,\nu)}{\partial r}$$
$$+ \frac{1-\mu^2}{r}\left[1 + \frac{\mu V(r)}{c}\left(1 - \frac{d\ln V}{d\ln r}\right)\right]\frac{\partial I(r,\mu,\nu)}{\partial \mu}$$
$$- \frac{\nu V(r)}{cr}\left[(1-\mu^2) + \mu^2\frac{d\ln V}{d\ln r}\right]\frac{\partial I(r,\mu,\nu)}{\partial \nu}$$
$$+ \frac{3V(r)}{cr}\left[(1-\mu^2) + \mu^2\frac{d\ln V}{d\ln r}\right]I(r,\mu,\nu)$$
$$= \eta(r,\nu) - \chi(r,\nu)I(r,\mu,\nu). \tag{6.2.1}$$

In (6.2.1), we have dropped $\partial/\partial t$ term from (2.5.10) as we are considering steady state. We have also dropped the sub- and superscripts zero to indicate radiation and physical variables in comoving frame. As all the variables in (6.2.1) are in comoving frame, there is no scope for confusion.

Now comparing (6.2.1) with (6.1.1) we note that the second terms in the coefficients of $\partial I/\partial r$ and $\partial I/\partial \mu$ are new. These terms arise from the physical effects of "advection" and "aberration". The complex structure of the space derivative d/ds on the left hand side of the transfer equation puts some limit to the direct differencing of equation (6.2.1). However, the spatial derivative terms are linear. Hence when the nature of the velocity field is given, one can find the characteristics of the paths of rays along which the spatial operator is a perfect differential.

The equations describing the characteristics $[r, (s,p), \mu(s,p)]$ can be written as [cf. (6.2.1)]

$$\frac{dr}{ds} = \mu + \frac{V(r)}{c} \tag{6.2.2}$$

and

$$\frac{d\mu}{ds} = \frac{1-\mu^2}{r}\left[1 + \frac{\mu V(r)}{c}\left(1 - \frac{d\ln V}{d\ln r}\right)\right]. \tag{6.2.3}$$

In $[r(s,p), \mu(s,p)]$, p is the closest distance of the trajectory from $r = 0$ and is the impact parameter and s is the distance along the ray from a chosen origin such that

$$\frac{dI}{ds} = \frac{\partial I}{\partial r}\frac{dr}{ds} + \frac{\partial I}{\partial \mu}\frac{d\mu}{ds}. \tag{6.2.4}$$

Our aim is to determine the radiation field on the radial mesh. The characteristics are chosen tangents to a set of spherical shells, the point of tangency is defined by $s = 0$ and for each discrete value of r, we put

$$r = p, \quad \frac{dr}{ds} = 0, \quad s = 0. \tag{6.2.5}$$

Then from (6.2.2)

$$\mu(s=0) = -V(r=p)/c. \tag{6.2.6}$$

Then treating (6.2.5) and (6.2.6) as initial conditions, (6.2.2) and (6.2.3) are integrated from $s = 0$ in the positive and negative directions to obtain the trajectories $[r(s), \mu(s)]$ for $V(r) \neq 0$.

Along the characteristics of the rays defined as above, the transfer equation can be written as [cf. (6.2.1), (6.2.2)–(6.2.4)]

$$\frac{\partial I}{\partial s}(s, p, \nu) - \frac{\tilde{a}}{r}[(1 - \mu^2) + b\mu^2]\frac{\partial}{\partial \nu}I(s, p, \nu)$$
$$= \eta(s, p, \nu) - \chi(s, p, \nu)I(s, p, \nu)$$
$$- \frac{3\tilde{a}}{\nu r}[(1 - \mu^2) + b\mu^2]I(s, p, \nu), \tag{6.2.7}$$

where

$$\tilde{a} = \nu V(r)/c \quad \text{and} \quad b = (d \ln V)/(d \ln r) \tag{6.2.8}$$

$\tilde{a}, r, \mu$ are functions of s and p.

The equation (6.2.7) differs from (6.1.15) in two respects. In it an additional term occurs on the right hand side and secondly in this case, the characteristic rays are not symmetric about the centre of the sphere. As a consequence, it is not possible to introduce mean intensity like and flux like variables, u and v here and the equation (6.2.7) cannot be transformed into second order form as is done in the Feautrier scheme. This difficulty is avoided in the following way.

The spatial and frequency-impact parameter discretisation is made as follows using the grids

$\{s_d\}$: spatial grid, $d = 1, 2, \ldots, D$
$\{p_i\}$: impact parameter grid, $i = 1, 2, \ldots, I$
$\{\nu_{k=n,i}\}$: frequency-impact parameter grid, $k = 1, 2, \ldots, NI$.

6.2 The Role of Aberration and Advection

We now write down (6.2.7) in the form of a finite difference equation

$$\left[\left(I_{(d+1/2),k} - I_{d,k}\right) \Big/ \triangle s_{(d+1/2)}\right]$$
$$-\bar{\gamma}_{(d+1/2),(k-1/2)}\left[I_{(d+1/2),(k-1)} - I_{(d+1/2),k}\right]$$
$$= \eta_{(d+1/2),k} - \chi^*_{(d+1/2),k} I_{(d+1/2),k}. \tag{6.2.9}$$

In (6.2.9), $\bar{\gamma}(s,p)$ is defined in (6.1.17). χ^* includes the second and the third terms of the left hand side of (6.2.7).

The numerical integration of first order linear equation is beset with the same difficulty due to the presence of a disturbing exponentially increasing solution. This is avoided by obtaining a formal analytical solution which is then discretised.

To do this a sort of optical depth and source function is introduced at every frequency impact parameter mesh point k given by

$$dt_k = (\chi^*_k + \bar{\gamma}_{k-1/2})ds \tag{6.2.10}$$

and

$$\left. \begin{array}{l} S^*_k = (\eta_k - \bar{\gamma}_{k-1}I_{k-1})/(\chi^*_k + \bar{\gamma}_{k-1/2}) \\ = \xi S_k + \vee_k I_{k-1} \end{array} \right\} \tag{6.2.11}$$

where S^*_k, the source function for two level atoms can be written as

$$\xi_k S^*_k = \xi_k \mathbf{J} + \Psi_k. \tag{6.2.12}$$

The equation (6.2.7) can be written as

$$\frac{dI_k}{dt_k} = -I_k + S^*_k \tag{6.2.13}$$

we now discretise I_k and S^*_k on the depth mesh $\{s_d\}$.

$$\mathbf{I}_k = (I_{1,k}, I_{2,k}, \ldots, I_{D,k}) \tag{6.2.14}$$

and

$$\mathbf{S}^*_k = (S^*_{1,k}, S^*_{2,k}, \ldots, S^*_{D,k}). \tag{6.2.15}$$

The formal solution of (6.2.13) can be written as

$$\mathbf{I}_k = \mathbb{L}_k \mathbf{S}^*_k + \mathbf{E}_k, \tag{6.2.16}$$

where the elements of matrix are evaluated analytically in terms of exponentials, when S_k^* is given.

From (6.2.11) and (6.2.12), we have

$$S_k^* = \mathbb{Z}_k \mathbf{J} + \mathbf{\Psi}_k + \mathbb{O}_k \mathbf{I}_{k-1}, \qquad (6.2.17)$$

where we can discretise $\mathbf{J}$ as

$$\mathbf{J} = (J_1, J_2, \ldots, J_D). \qquad (6.2.18)$$

In (6.2.17) $\mathbb{Z}_k$ is a diagonal and $\mathbb{O}_k$ is a Chevron matrix.

The frequency elimination procedure employed here is the same as in section (6.1). We obtain I_k of the form

$$\mathbf{I}_k = \mathbb{A}_k \mathbf{J} + \mathbf{B}_k \qquad (6.2.19)$$

where

$$\mathbb{A}_k = \mathbb{L}_k [\mathbb{Z}_k + \mathbb{O}_k \mathbb{A}_{k-1}] \qquad (6.2.20)$$
$$\mathbf{B}_k = \mathbb{L}_k [\mathbf{\Psi}_k + \mathbb{O}_k \mathbf{B}_{k-1}] + \mathbf{E}_k. \qquad (6.2.21)$$

For starting the recursion scheme, one takes $\mathbb{O}_1 = 0$ at the further wing which is the continuum. Writing

$$J_j = \sum_k w_k I_{k,j}, \qquad (6.2.22)$$

where w_k is the quadrature weight, and substituting (6.2.19) for all frequencies and impact parameters in (6.2.22), we get the final system for $\mathbf{J}$ as

$$\mathbb{C} \mathbf{J} = \mathbf{D} \qquad (6.2.23)$$

which is solved for $\mathbf{J}$.

The scheme of solution presented above is available for more general form of transfer equations in comoving frame, for example those involving higher powers of V/c or relativistic form of transfer equation in comoving frame.

For $(V/c) \ll 1$, which holds good in most of stellar wind problems Michalas, Kunasz and Hummer (1976) have developed a perturbation method of solution, which reduces the problem to one of formal similarity to that in section (6.1). Through some examples, they concluded that contributions from advection and aberration terms can be ignored in comparison with that of Doppler effect, in studying the transfer problems in spherically symmetric expanding atmospheres. However, in the case of supernovae or galactic winds with large velocity, advection and aberration may make significant contributions.

References

Castor, J.I. (1972): Ap. J., **178**, 439.
Castor, J.I., (1986): Problems in Astrophysical Radiation Hydrodynamics, ed. by Winkler and Norman, Reidel Publishing Co., p. 1.
Kunasz, P.B., Hummer, D.G. (1974): MNRAS, **166**, 57.
Mihalas, D., Kunasz, P.B., Hummer, D.G. (1975): Ap. J., **202**, 465.
Mihalas, D., Kunasz, P.B., Hummer, D.G. (1976): Ap. J., **203**, 467.
Mihalas, D., Kunasz, P.B., Hummer, D.G. (1976): Ap.J., **206**, 505.
Mihalas, D., Kunasz, P.B., Hummer, D.G. (1976): Ap. J., **210**, 419.
Mihalas, D., Kunasz, P.B. (1978): Ap. J., **217**, 635.
Mihalas, D. (1978): Stellar Atmospheres. W. Freeman and Co., San Francisco.
Mihalas D. (1980): Ap. J., **237**, 574.
Mihalas, D., Mihalas, B.W. (1984): Foundations of Radiation Hydrodynamics, Oxford University Press, New York.
Mihalas, D. (1986): Problems in Astrophysical Radiation Hydrodynamics, ed. by Winkler and Norman. Reidel Publishing Co., p. 45.

Chapter 7. Moment Equations in Comoving Frame and Their Solutions

In chapter 6, the radiative transfer equation is spherically symmetric moving media was solved by the impact parameter method. Quite often Rybicki's elimination scheme was preferred to that of Feautrier's to save the cost of computation. In Rybicki's scheme, a complete redistribution of photons in scattering by atoms and electrons was assumed. Thus the source function for each source becomes frequency-independent and isotropic. Hearn (1964), Hummer (1969), Milkey and Mihalas (1973) compared the solutions of transfer equation under complete frequency redistribution with those of Hummer (1962), where the scattering integral was expressed in terms of angle-averaged redistribution function. They dealt with static atmospheres and concluded that as long as the extreme wings of the line participate in transfer, complete redistribution of frequency is a fairly good approximation. However, for a moving medium, the isotropy of the source function, its independence on frequency and the symmetry of the line profile about the line centre cannot always be taken for granted. The wisdom of the use of complete redistribution function in such cases is suspect. Morgan (1974) and Mihalas *et al.* (1976) examined the problem and concluded that the complete redistribution function and angle-averaged redistribution function in the comoving frame practically produce similar line profile.

In this chapter, we shall review the work of Mihalas *et al.* (1976), where the radiative transfer in comoving frame in spherical geometry has been solved with an arbitrary angle-averaged redistribution function. The moment equations in comoving frame have been taken with the introduction of the variable Eddington factor to eliminate the angle-dependence of intensity. Feautrier (1964) scheme of elimination has been used. The comoving frame moment equations in spherical geometry involving special relativistic flows have also been solved. A short review of the use of generalised Eddington approximation in solving such a problem has also been given.

7.1 The Moment Equations and Their Solutions

The method outlined below was proposed by Mihalas et al. (1976). They started with the moment form of transfer equation for a spherically symmetric atmosphere with a radial flow of velocity $V(r)$. In comoving frame, the radiative transfer equation for radiation of frequency ν propagating in the direction $\cos^{-1}\mu$ with the radius vector r as recorded by an observer comoving with the element of the atmosphere at r is given in the time-independent case as [cf. (2.1.14)]

$$\mu\frac{\partial I(r,\mu,\nu)}{\partial r} + \frac{1-\mu^2}{r}\frac{\partial I(r,\mu,\nu)}{\partial \mu}$$
$$-\frac{\nu V(r)}{cr}\left[1-\mu^2+\mu^2\left(\frac{d\ln V(r)}{d\ln r}\right)\right]\frac{\partial I(\mu,\nu,r)}{\partial \nu}$$
$$= \eta_\nu - \chi_\nu I(r,\mu,\nu). \tag{7.1.1}$$

In (7.1.1) all the variables are referred to the comoving frame. The subscript and superscript zero used to label them in (2.1.14) have been dropped. The effects of advection and aberration are considered negligible compared to the contribution from Doppler effect. The symbols in (7.1.1) have their usual meanings.

We introduce the angular-moments of the specific intensity in the following

$$M_{\nu n} = \frac{1}{2}\int_{-1}^{1} I(r,\mu,\nu)\mu^n d\mu, \quad n=0,1,2,\ldots. \tag{7.1.2}$$

Thus

$$M_{\nu 0} = \frac{1}{2}\int_{-1}^{1} I(r,\mu,\nu)d\mu = J_\nu(r,\nu) \tag{7.1.3}$$

$$M_{\nu 1} = \frac{1}{2}\int_{-1}^{+1} I(r,\mu,\nu)\mu d\mu = H_\nu(r,\nu) \tag{7.1.4}$$

7.1 The Moment Equations and Their Solutions

$$M_{\nu 2} = \frac{1}{2}\int_{-1}^{+1} I(r,\mu,\nu)\mu^2 d\mu = K_\nu(r,\nu) \tag{7.1.5}$$

$$M_{\nu 3} = \frac{1}{2}\int_{-1}^{+1} I(r,\mu,\nu)\mu^3 d\mu = L_\nu(r,\nu) \tag{7.1.6}$$

and so on, we also write

$$\frac{\nu V(r)}{cr} = \tilde{a}, \qquad \frac{d\ln V(r)}{d\ln r} = b, \tag{7.1.7}$$

the Eddington factors f_ν and g_ν are given by

$$f_\nu = K_\nu/J_\nu \text{ and } g_\nu = L_\nu/H_\nu, \tag{7.1.8}$$

where $f_\nu = f_\nu(r,\nu)$ and $g_\nu = g_\nu(r,\nu)$.

Then the first two angular moment equations corresponding to (7.1.1) are

$$\frac{1}{r^2}\frac{\partial}{\partial r}(r^2 H_\nu) - \tilde{a}\left[\frac{\partial}{\partial \nu}\{(1-f_\nu)J_\nu\} + b\frac{\partial}{\partial \nu}(f_\nu J_\nu)\right] = \eta_\nu - \chi_\nu J_\nu \tag{7.1.9}$$

and

$$\frac{\partial}{\partial r}(f_\nu J_\nu) + \frac{(3f_\nu - 1)}{r}J_\nu$$
$$-\tilde{a}\left[\frac{\partial}{\partial \nu}\{(1-g_\nu)H_\nu\} + b\frac{\partial}{\partial \nu}(f_\nu J_\nu)\right]$$
$$= -\chi_\nu H_\nu. \tag{7.1.10}$$

We further introduce q_ν through (Auer (1971))

$$\frac{\partial}{\partial r}(r^2 q_\nu) = \frac{3f_\nu - 1}{f_\nu r}, \tag{7.1.11}$$

and write

$$dX_\nu = -\chi_\nu q_\nu dr \tag{7.1.12}$$
$$\tilde{a}/\chi_\nu = \gamma_\nu. \tag{7.1.13}$$

Then the transfer equations (7.1.9) and (7.1.10) can be rewritten as

$$q_\nu \frac{\partial}{\partial X_\nu}(r^2 H_\nu) + \gamma_\nu\left[\frac{\partial}{\partial \nu}\{(1-f_\nu)r^2 J_\nu\} + b(f_\nu r^2 J_\nu)\right]$$
$$= r^2(J_\nu - S_\nu) \tag{7.1.14}$$

$$\frac{\partial}{\partial X_\nu}(f_\nu q_\nu r^2 J_\nu) + \gamma_\nu \left[\frac{\partial}{\partial \nu}\{(1-g_\nu)r^2 H_\nu\} + b\frac{\partial}{\partial \nu}(g_\nu r^2 H_\nu)\right]$$
$$= r^2 H_\nu. \tag{7.1.15}$$

The next step is to discretise the variables and write the equations (7.1.14) and (7.1.15) in the finite difference equation form. The variables are discretised on a depth-frequency grid.

$$\{r_d\}: \quad \text{Depth points}, \quad d = 1, 2, \ldots, D$$
$$\{\nu_k\}: \quad \text{frequency points}, \quad k = 1, 2, \ldots, K.$$

After Mihalas et al, the index of the depth grid is taken to increase from the outer surface inwards and the sign of the frequency depends on the relative signs of $V(r)$ and $\frac{dV}{dr}$. The frequencies are measured from blue to red.

The equation (7.1.15) written as a finite difference equation reads as

$$\frac{f_{(d+1),k} q_{(d+1),k} r^2_{(d+1)} J_{(d+1),k} - f_{d,k} q_{d,k} r^2_d J_{d,k}}{\Delta X_{(d+1/2),k}}$$
$$+ \frac{\gamma_{(d+1/2)(k-1/2)} r^2_{(d+1/2)}}{\Delta \nu_{k-1/2}} \Big[(1 - g_{(d+1/2),(k-1)})$$
$$+ b_{(d+1/2)} g_{(d+1/2),(k-1)}) H_{(d+1/2),(k-1)}$$
$$- (1 - g_{(d+1/2),k} + b_{(d+1/2)} g_{(d+1/2),k}) H_{(d+1/2),k}\Big]$$
$$= r^2_{(d+1/2)} H_{(d+1/2),k}, \tag{7.1.16}$$

where
$$\Delta \nu_{k-1/2} = (\nu_{k-1} - \nu_k) \tag{7.1.17}$$
$$r_{d+1/2} = \frac{1}{2}[r_d + r_{d+1}] \tag{7.1.18}$$

$$\left.\begin{array}{rl} \tilde{a}_{d+1/2} &= \tilde{a}_{(r_{d+1/2})} \\ b_{d+1/2} &= b(r_{d+1/2}) \\ g_{(d+1/2),k} &= g(r_{d+1/2}, \nu_k) \end{array}\right\} \tag{7.1.19}$$

$$\gamma_{(d+1/2),(k-1/2)} = \tilde{a}_{d+1/2}/\chi_{(d+1/2),(k-1/2)} \tag{7.1.20}$$

7.1 The Moment Equations and Their Solutions

$$\chi_{(d+1/2),(k-1/2)} = \frac{1}{4}[\chi_{(d+1),k} + \chi_{d,k} + \chi_{(d+1),(k-1)} + \chi_{d,(k-1)}]. \quad (7.1.21)$$

The finite difference representation (7.1.16) of (7.1.15) can be rewritten as

$$r_{d+1/2}^2 H_{(d+1/2),k} = \frac{f_{(d+1),k} q_{(d+1),k} r_{(d+1)}^2 J_{(d+1),k} - f_{d,k} q_{d,k} r_d^2 J_{d,k}}{(1 + w_{(d+1/2),k,k}) \Delta X_{(d+1/2),k}}$$
$$+ \frac{w_{(d+1/2),k,(k-1)}}{1 + w_{(d+1/2),k,k}} r_{(d+1/2)}^2 H_{(d+1/2),(k-1)} \quad (7.1.22)$$

where

$$w_{(d+1/2),k,\ell} = \frac{\gamma_{(d+1/2),(k-1/2)} r_{d+1/2}^2}{\Delta \nu_{k-1/2}} (1 - g_{(d+1/2),\ell} + b_{(d+1/2)} g_{(d+1/2),\ell}). \quad (7.1.23)$$

At the extreme line wing, i.e. at the bluest frequency line merges to continuum radiation, which can be taken to be frequency independent. Then at the extreme line wing, we may assume $(\partial I/\partial \nu) = 0$. This implies

$$w_{(d+1/2),k,1} = 0. \quad (7.1.24)$$

The equation (7.1.22) together with (7.1.24) allows one to develop a scheme for obtaining a recursive solution in k for $r_{d+1/2}^2 H_{(d+1/2),k}$ and we have

$$r_{d+1/2}^2 H_{(d+1/2),k}$$
$$= \sum_{k'=1}^{k} \Psi_{(d+1/2),k,k'}$$
$$\times \left(\frac{f_{(d+1),k'} q_{(d+1),k'} r_{d+1}^2 J_{(d+1),k'} - f_{d,k'} q_{d,k'} r_d^2 J_{d,k'}}{\Delta X_{(d+1/2),k'}} \right) \quad (7.1.25)$$

where

$$\Psi_{(d+1/2),k,k'} = \frac{1}{1 + w_{(d+1/2),k,k}} \prod_{\ell=k'+1}^{k} \frac{w_{(d+1/2),\ell,\ell-1}}{1 + w_{(d+1/2),\ell,\ell}}. \quad (7.1.26)$$

The product is taken to be unity when $k = k'$.

Similarly discretising the variables in radius-frequency grid and writing (7.1.14) as a finite difference equation, we have

$$\frac{q_{d,k}}{\Delta X_{d,k}}[r_{d+1/2}^2 H_{(d+1/2),k} - r_{d-1/2}^2 H_{(d-1/2),k}]$$
$$+\frac{\gamma_{d,(k-1)}}{\Delta \nu_{k-1/2}} \left[(1 - f_{d,(k-1)} + b_d f_{d,(k-1)})r_d^2 J_{d,(k-1)}\right.$$
$$\left. -(1 - f_{d,k} + b_d f_{d,k})r_d^2 J_{d,k}\right]$$
$$= r_d^2[J_{d,k} - S_{d,k}], \qquad (7.1.27)$$

where $S_{d,k}$ is given by

$$S_{d,k} = \sum_{k'} w_{k'} \mathcal{R}_{d,k',k} J_{d,k'} + \varepsilon_{d,k}. \qquad (7.1.28)$$

(7.1.28) is the finite difference form of the expression for source function given by

$$S(r,\nu) = \xi(r,\nu) \int \mathcal{R}(r,\nu',\nu) J(r,\nu') d\nu' + \varepsilon(r,\nu) \qquad (7.1.29)$$

where ξ and ε are known functions of radius and frequency. $\mathcal{R}(r,\nu',\nu)$ is an angle-averaged redistribution function depending on radius.

From (7.1.27), we eliminate $H_{(d+1/2),k}$ by (7.1.25) and get

$$\frac{q_{d,k}}{\Delta X_{d,k}} \left[\sum_{k'=1}^{k} \frac{\Psi_{(d+1/2),k,k'}}{\Delta X_{(d+1/2),k}} \left\{ f_{(d+1),k'} q_{(d+1),k'} r_{(d+1)}^2 J_{(d+1),k'} \right. \right.$$
$$\left. - f_{d,k'} q_{d,k'} r_d^2 J_{d,k'} \right\}$$
$$- \sum_{k'=1}^{k} \frac{\Psi_{(d-1/2),k,k'}}{\Delta X_{(d-1/2),k'}} \left\{ f_{d,k'} q_{d,k'} r_d^2 J_{d,k'} \right.$$
$$\left. \left. - f_{(d-1),k'} q_{(d-1),k'} r_{d-1}^2 J_{(d-1),k'} \right\} \right]$$
$$= r_d^2 \left[(1 + \Pi_{d,k,k}) J_{d,k} - S_{d,k} - \Pi_{d,k,(k-1)} J_{d,(k-1)} \right] \qquad (7.1.30)$$

where

$$\Pi_{d,k,\ell} = \frac{\gamma_{d,(k-1/2)}}{\Delta \nu_{k-1/2}} (1 - f_{d,\ell} - b f_{d,\ell}). \qquad (7.1.31)$$

The initial condition at the extreme wing is

7.1 The Moment Equations and Their Solutions

$$\Pi_{d,1,\ell} = 0. \tag{7.1.32}$$

The Boundary Conditions

(a) At the upper boundary of the atmosphere $I(0, \nu_k) = 0$ which implies

$$X_{1,k} = 0. \tag{7.1.33}$$

(b) At the lower boundary $X_{D,k}$, which is deep in the atmosphere, the diffusion approximation is supposed to hold good. That is at the lower boundary,

$$H_\nu = (1/3\chi_\nu) \left| \frac{\partial B_\nu}{\partial r} \right|_{r_0}, \quad r_0 = r_D. \tag{7.1.34}$$

These two boundary conditions are put in the difference form.

To deal with the upper boundary condition we introduce two new Eddington factors at the upper boundary $r = R(X_{1,k} = 0)$. These are

$$h_k = H_{1,k}/J_{1,k} = \frac{\int_0^1 I(R, \mu, \nu_k)\mu d\mu}{\int_0^1 I(R, \mu, \nu_k)d\mu} \tag{7.1.35}$$

$$\ell_k = L_{1,k}/J_{1,k} = \frac{\int_0^1 I(R, \mu, \nu_k)\mu^3 d\mu}{\int_0^1 I(R, \mu, \nu_k)d\mu}. \tag{7.1.36}$$

Now writing (7.1.16) for $d = 1$, taking (7.1.35) and (7.1.36) into account, we have

$$\frac{f_{2,k}q_{2,k}r_2^2 J_{2,k} - f_{1,k}q_{1,k}r_1^2 J_{1,k}}{\Delta X_{3/2,k}}$$
$$= r_1^2[(h_k + \theta_{1,k,k})J_{1,k} - \theta_{1,k,(k-1)}J_{1,(k-1)}], \tag{7.1.37}$$

where

$$\theta_{1,k,\ell} = \frac{\gamma_{1,k-1/2}}{\Delta \nu_{k-1/2}}(h_\ell - \ell_\ell + b\ell_\ell) \tag{7.1.38}$$

$$\gamma_{1,(k-1/2)} = \tilde{a}(R)/\chi_{1,(k-1/2)} \tag{7.1.39}$$

$$\chi_{1,(k-1/2)} = \frac{1}{2}[\chi_{1,k} + \chi_{1,k-1}]. \qquad (7.1.40)$$

The boundary condition (b) at the lower boundary $d = D$ written in the form of finite difference equation reads as

$$\frac{f_{D,k} q_{D,k} r_D^2 J_{D,k} - f_{(D-1),k} q_{(d-1),k} r_{(d-1)}^2 J_{(D-1),k}}{\Delta X_{(D-1/2),k}}$$

$$= \frac{r_0^2}{3\chi_{D,k}} \left| \frac{\partial B(\nu_k)}{\partial r} \right|_{r=r_0} \qquad (7.1.41)$$

where $r_0 = r_D$.

Elimination Scheme

For the solution of equation (7.1.30) under the boundary conditions (7.1.37), (7.1.41), where all of them have been put in a finite difference equation form, we use the Gaussian Elimination scheme of Feautrier (1964).

We introduce the vector $\mathbf{J}_d$ given by

$$\mathbf{J}_d \equiv (J_{d,1}, J_{d,2}, \ldots, J_{d,k}). \qquad (7.1.42)$$

The radiative transfer equation (7.1.30) and the boundary conditions (7.1.37) and (7.1.41) can be combined and written in the matrix form

$$-\mathbb{A}_d \mathbf{J}_{(d-1)} + \mathbb{B}_d \mathbf{J}_d - \mathbb{C}_d \mathbf{J}_{(d+1)} = \mathbf{L}_d, \qquad (7.1.43)$$

where $d = 1, 2, \ldots, D$.

In (7.1.43), $\mathbb{B}_d$ is a full matrix and $\mathbb{A}_d$ and $\mathbb{C}_d$ are triangular matrices of dimension $K \times K$.

The complete system of equations (7.1.43) are solved ray-by-ray, frequency by frequency, the initial condition being kept in mind. $f_{d,k}, g_{d,k}, h_k, \ell_k$ are taken to be assigned parameters. Hence from (7.1.43) one can determine $J_{d,k}$ and $S_{d,k}$ for all depths and frequencies following the basic steps of Feautrier's elimination scheme as described in equations (6.1.44)–(6.1.57). $S_{d,k}$ being known, the transfer equations in comoving frame is formally integrated along a number of rays labelled by appropriate impact parameter to obtain the radiation field frequency by frequency.

Mihalas, Kunasz and Hummer (1976) solved the line transfer problems for some specific atmospheric models for both complete and angle averaged redistribution functions. The calculations were made with different velocity laws of motion in atmospheric models. Mihalas and Kunasz (1978) applied the methods developed in this and the previous chapter to the solution of coupled transfer and statistical equilibrium equations for multilevel atoms in model atmospheres. The equations were solved iteratively using equivalent two level atom approach. A complete linearisation scheme was formulated to take care of situations where equivalent two level atom approach failed.

7.2 Comoving Frame Transfer Equations in Relativistic Flows

In section (6.1), we have considered the radiative transfer problem in a spherically symmetric moving medium writing the transfer equation in a comoving frame where only the Doppler shift was taken into account. In section (6.2) we studied the importance of including the aberration and advection terms in the radiative transfer equation in comoving frame. Attention was also drawn to the fact the contributions of Doppler effect, aberration and advection are all of first order in V/c. In section (7.1), the moment equations were set in a comoving frame and solved only when the Doppler shift is included for consideration. However, in section (6.2) it was concluded that for small velocities [$(V/c \simeq 0.01]$, the contributions from Doppler shift outweighs those from aberration and advection. But when (V/c) is not too small and gradually approaches unity, all the terms of the order of (V/c) should be retained. For high velocities, the escape probability of photons increases and emitted photons hardly suffer more than single scattering. Under these circumstances, the problem of continuum formation supersedes that of line formation. It is thought useful to write out the radiative transfer equation in comoving frame in more general form.

In what follows we shall write down the less restricted form of radiative transfer equation than before, under the frame work

7. Moment Equations in Comoving Frame

of special relativity and try to solve it for spherically symmetric moving medium. The moment equations will be considered in comoving frame [cf. Mihalas (1980)].

The Equations of Transfer

In chapter 2, we have expressed the radiative transfer equations for spherically symmetric flows in lab, comoving and mixed frames. We have also deduced the monochromatic and frequency integrated moment equations in the comoving frame and the relevant energy and momentum equations.

The general form of transfer equation for spherically symmetric flow within the scope of relativistic frame work is given by [cf. equation (2.5.18)]

$$\frac{\gamma}{c}(1+\beta\mu_0)\frac{\partial I^0(\mu_0,\nu_0)}{\partial t} + \gamma(\mu_0+\beta)\frac{\partial I^0(\mu_0,\nu_0)}{\partial r}$$
$$+\frac{\partial}{\partial \mu_0}\left[\gamma(1-\mu_0^2)\left\{\frac{(1+\beta\mu_0)}{r} - \gamma^2(\mu_0+\beta)\frac{\partial \beta}{\partial r}\right.\right.$$
$$\left.\left. -\frac{\gamma^2}{c}(1+\beta\mu_0)\frac{\partial \beta}{\partial t}\right\} I^0(\mu_0,\nu_0)\right] - \frac{\partial}{\partial \nu_0}\left[\gamma\nu_0\left\{\beta\frac{(1-\mu_0^2)}{r}\right.\right.$$
$$\left.\left. +\gamma^2\mu_0(\mu_0+\beta)\frac{\partial \beta}{\partial r} + \frac{\gamma^2}{c}\mu_0(1+\beta\mu_0)\frac{\partial \beta}{\partial t}\right\} I^0(\mu_0,\nu_0)\right]$$
$$+\gamma\left[\frac{2\mu_0+\beta(3-\mu_0^2)}{r} + \gamma^2(1+\mu_0^2+2\beta\mu_0)\frac{\partial \beta}{\partial r}\right.$$
$$\left. +\frac{\gamma^2}{c}\{2\mu_0+\beta(1+\mu_0^2)\}\frac{\partial \beta}{\partial t}\right] I^0(\mu_0,\nu_0)$$
$$= \eta^0(\nu_0) - \chi^0(\nu_0)I^0(\mu_0,\nu_0) \qquad (7.2.1)$$

where $I^0(\mu_0,\nu_0) \equiv I^0(r,t;\mu_0,\nu_0)$, $\gamma = (1-\beta^2)^{-1/2}$ and $\beta = V/c$. In (7.2.1), the sub- and superscripts zero to the variables indicate that they are being measured in comoving frame. As all the radiation and material quantities in (7.2.1) are measured in comoving frame, we shall recognize that (7.2.1) is a comoving frame equation and drop the sub and superscripts to label the variables in the subsequent part of the chapter.

Now we introduce the angular moments,

$$[J(\nu), H(\nu), K(\nu), L(\nu)] = \frac{1}{2}\int_{-1}^{1} I(\mu_0, \nu_0)[\mu^0, \mu^1, \mu^2, \mu^3]d\mu. \tag{7.2.2}$$

Then integrating the equation (7.2.1) over angles (for successive powers of μ) and frequencies in the comoving frame we obtain the frequency integrated moment equations.

For the zeroth moment, we have

$$\frac{\gamma}{c}\left[\frac{\partial J}{\partial t} + \beta\frac{\partial H}{\partial t}\right] + \gamma\left(\frac{\partial H}{\partial r} + \beta\frac{\partial J}{\partial r}\right)$$
$$+\gamma\left[\frac{1}{r}(2H + 3\beta J - \beta K) + \frac{\gamma^2}{c}\frac{\partial \beta}{\partial t}(2H + \beta J + \beta K)\right]$$
$$+\gamma^2\frac{\partial \beta}{\partial r}(J + K + 2\beta H)$$
$$= q_0. \tag{7.2.3}$$

For the first moment, we have

$$\frac{\gamma}{c}\left(\frac{\partial H}{\partial t} + \beta\frac{\partial K}{\partial t}\right) + \gamma\left(\frac{\partial K}{\partial r} + \beta\frac{\partial H}{\partial r}\right)$$
$$+\gamma\left[\frac{1}{r}(3K - J + 2\beta H) + \frac{\gamma^2}{c}\frac{\partial \beta}{\partial t}(J + K + 2\beta H)\right]$$
$$+\gamma^2\frac{\partial \beta}{\partial r}(2H + \beta J + \beta K)$$
$$= q_1 \tag{7.2.4}$$

where

$$q_n = \frac{1}{2}\int_0^\infty d\nu \int_{-1}^{+1}[\eta(\nu) - \chi(\nu)]\mu^n d\mu \tag{7.2.5}$$

with $n = 0, 1, 2, \ldots$.

Equations (7.2.3) and (7.2.4) are frequency integrated zeroth and first moment equations for special relativistic radiation flows through spherically symmetric atmosphere.

A close look at the equations (7.2.1)–(7.2.5) reveals that the moment equations are pretty complicated and less amenable to conversion into difference equations of the type considered so far. This is true even in the case of steady flow, where we suppress all terms involving time derivatives. The archetype of these steady flow conditions obtain in uniform non-accelerating expansions of medium and steady stellar winds. Moreover, the moment equations are invariably attended with search for appropriate closure conditions and these are hard to secure without reference to the angle and frequency dependent equations of transfer. So in the present problem, the conversion of radiative transfer equations to frequency-integrated moment form does not give any additional advantage in the solution of the problem. For simplicity, we shall drop all the time dependent terms in (7.2.2) as their inclusion is not expected to bring about any basic change in the mathematical methodology to be used here.

Thus the basic radiative transfer equation for special relativistic steady flow of radiation in a spherically symmetric atmosphere in comoving frame reads as follows.

$$\gamma(\mu+\beta)\frac{\partial I(\mu,\nu)}{\partial r} + \gamma(1-\mu^2)\left[\frac{1+\beta\mu}{r} - \gamma^2(\mu+\beta)\frac{\partial \beta}{\partial r}\right]$$
$$\times \frac{\partial I(\mu,\nu)}{\partial \mu} - \gamma\nu\left[\frac{\beta(1-\mu^2)}{r} + \gamma^2\mu(\mu+\beta)\frac{\partial \beta}{\partial r}\right]\frac{\partial I(\mu,\nu)}{\partial \nu}$$
$$+3\gamma\left[\frac{\beta(1-\mu^2)}{r} + \gamma^2\mu(\mu+\beta)\frac{\partial \beta}{\partial r}\right]I(\mu,\nu)$$
$$= \eta(\nu) - \chi(\nu)I(\mu,\nu). \tag{7.2.6}$$

In (7.2.6), all the quantities are measured in comoving frame.

In solving (7.2.6), we meet with a situation similar to that in (6.2.1). The complicated nature of the structure of space derivative term in the left hand side of the equation puts some limitation to the direct expression of (7.2.6) as a finite difference equation. However, one can draw the characteristics of trajectory of each ray along which the spatial operator is a perfect differential. The specific impact parameter p is chosen for each ray. p is the closest distance of the trajectory from $r = 0$. Let s be the distance along the ray from a chosen origin.

The trajectories are determined by $[r(s,p), \mu(s,p)]$ such that

$$\frac{dI}{ds} = \frac{\partial I}{\partial r}\frac{dr}{ds} + \frac{\partial I}{\partial \mu}\frac{d\mu}{ds}. \tag{7.2.7}$$

The characteristic equations are given by

$$\frac{dr}{ds} = \gamma(\mu + \beta) \tag{7.2.8}$$

$$\frac{d\mu}{ds} = \gamma(1-\mu^2)\left[\frac{1+\beta\mu}{r} - \gamma^2(\mu+\beta)\frac{\partial\beta}{\partial r}\right]. \tag{7.2.9}$$

For static media $\beta = 0$ and the characteristics are parallel straight lines.

Our aim in this calculation is to ascertain the radiation field on a chosen radial-mesh $\{r_d\}$. The characteristics are chosen as tangents to a set of spherical shells. The impact parameter p may be taken to be the radius r_d of one such shell. $s = 0$ denotes the origin of the path length and is the tangent point of the shell. Then at

$$s = 0 \quad \left.\frac{dr}{ds}\right|_{s=0} = 0. \tag{7.2.10}$$

Form (7.2.8), we find

$$\mu(s=0, p) = -\beta(p) \tag{7.2.11}$$

and from (7.2.9) and (7.2.11)

$$\left.\frac{d\mu}{ds}\right|_{s=0} = \frac{(1-\beta^2)^{3/2}}{r}. \tag{7.2.12}$$

The equations (7.2.10)–(7.2.12) are initial conditions for integrating equations (7.2.8) and (7.2.9) to determine the trajectories $[r(s,p), \mu(s,p)]$. The integration is done over both positive and negative s. For the nature of characteristics for linear and quadratic velocity laws and certain comments about the scheme of integration of (7.2.8) and (7.2.9) in the presence of an opaque stellar core of radius r_c, it is advisable to consult Mihalas [(1980) p. 581–582].

Solution of the Transfer Equation

We shall obtain a ray-by-ray solution of the problem as in section (6.2). For each value of p, the radiative transfer equation along each characteristic ray is given by

$$\frac{\partial I(s,\nu)}{\partial s} - a'(s,\nu)\frac{\partial I(s,\nu)}{\partial \nu} = \eta(s,\nu) - \chi'(s,\nu)I(s,\nu) \quad (7.2.13)$$

where

$$a'(s,\nu) = \gamma\nu\left[\frac{\beta(1-\mu^2)}{r} + \gamma^2\mu(\mu+\beta)\frac{\partial\beta}{\partial r}\right] \quad (7.2.14)$$

and

$$\chi'(s,\nu) = \chi(s,\nu) + 3\gamma\left[\frac{\beta(1-\mu^2)}{r} + \gamma^2\mu(\mu+\beta)\frac{\partial\beta}{\partial r}\right]. \quad (7.2.15)$$

Note should be taken of the fact that along a particular ray the frequency ν remains constant, but comoving frame angle $\cos^{-1}\mu$ changes. This change is taken care of by assigning adequate number of angle points to project the variation of the radiation field on each shell. The equation (7.2.13) has formal similarity with (6.2.8) in chapter 2. We shall tackle the problem here in the same way as we have done in section (6.2).

Along the set of characteristics, the spatial grid $\{r_d\}$ gives rise to a set of grids $\{s_d\}\{\mu_d\}$. The frequency grid is $\{\nu_k\}$. Then writing the frequency derivative in the finite-difference form in (7.2.13) we have

$$\frac{\partial I(s,\nu_k)}{\partial s} - \alpha_{k-1/2}(s)[I(s,\nu_{k-1}) - I(s,\nu_k)]$$
$$= \eta(s,\nu_k) - \chi'(s,\nu_k)I(s,\nu_k) \quad (7.2.16)$$

where

$$\alpha_{k-1/2}(s) = \frac{a'(s,\nu_k)}{\Delta\nu_{k-1/2}} = \frac{a'(s,\nu_k)}{\nu_{k-1} - \nu_k}. \quad (7.2.17)$$

We now take note of the fact that the characteristic rays are not symmetric about the origin in (7.2.14). This prevents us from introducing mean intensity and flux like variables as in Feautrier's scheme. The first order equation (7.2.14) will be required to be integrated over the entire length of the ray. For rays not intersecting the core, this length will be from surface to surface. For rays intercepted by the core, $(p < r_c)$ the integration will be surface to core and core to surface.

The equation (7.2.16) is discretised and written in the finite difference form. This equation turns out to be formally similar to equation (6.2.9) and its numerical equation is beset with the same difficulty due to the presence of exponentially increasing solution. It is avoided exactly in the same way as was done in section (6.2). This is done by obtaining a formal solution of (7.2.16) along each ray, the source function being represented by polynomials and splines on discrete mesh and performing the integration analytically. The process is followed for each frequency in the mesh $\{\nu_k\}$. For the solution of the set of equations a recursive method is followed [cf. equations (6.2.10)–(6.2.23)]. The boundary conditions considered are similar to those in section (6.2).

We state below the main features of the method.

(a) The transfer equation (7.2.16) is expressed as a finite difference equation in spatial and frequency discretised points $\{s_d\}$ and $\{\nu_k\}$.

(b) The numerical integration of the first-order differential equation is done at each frequency mesh point through a formal analytical solution. For this the transfer equation is written down in terms of specific intensity I_k and modified S_k^* (corresponding to χ').

(c) I_k and S_k^* are discretised on depth mesh $\{s_d\}$ and expressed as vectors.

$$\mathbf{I}_k \equiv [I_{1,k}, I_{2,k}, \ldots, I_{D,k}] \qquad (7.2.18)$$
$$\mathbf{S}_k^* \equiv [S_{1,k}^*, S_{2,k}^*, \ldots, S_{D,k}^*], \qquad (7.2.19)$$

and from the formal solution of the transfer equation $\mathbf{I}_k$ and $\mathbf{S}_k^*$ expressed through suitable matrix expressions.

(d) $\mathbf{J}$ involved in the expression of S_k^* is discretised and the frequency elimination scheme of Feautrier type is utilised. A recursion scheme is developed and final system for determining $\mathbf{J}$ obtained in the form

$$\mathbb{C}\mathbf{J} = \mathbf{D}, \qquad (7.2.20)$$

and it is solved for $\mathbf{J}$. Then knowing $\mathbf{J}$, we can know the other variables of the radiation field. The transfer equation for continuum can also be solved by some small modification and extension of this scheme [cf. Mihalas (1980) p. 583]. Mihalas (1980) also demonstrated the method for (i) grey radiative equilibrium problem and (ii) non-grey continuum problems.

7.3 Moment Methods Based on Generalised Eddington Approximation

Unno and Kondo (1976, 1977) proposed a generalisation of Eddington approximation for solving radiative transfer problems in static spherically symmetric medium. They introduced a variable $\mu_r(=\cos\theta_r)$ which permitted the radiation field to be divided into two distinct streams. $\mu_r(r,\nu)$ in turn was determined from the solution of the differential equation. The radiation field was described in terms of the first two moments J (zeroth) and F (first) of specific intensity. The second and the third moments were expressed in terms of J and F through the generalised Eddington relations. This method in a way evolved from the classical works of McCrea and Mitra (1936) and Chadrasekhar (1934) for solving transfer problems in spherical geometry. However, the introduction of unknown μ_r into the scheme and the idea of obtaining it from the solution of moment equations is a new concept and is credited to Unno and Kondo (1976, 1977). The basic features of Unno and Kondo's method are explained in Sen and Wilson (1990, p. 137–142).

Haisch (1979) applied this two-stream generalised Eddington approximation to tackle the radiative transfer problems in static, extended, spherically symmetric dust shell. He introduced the unknown parameter $\mu_r(r,\nu)$ for dividing the radiation field into two solid angular zones. He detailed the physical properties of the dust shells through opacity, emissivity etc and followed the procedure of Unno and Kondo to solve the transfer problems in a thick dust shell. He proposed a novel iterative scheme for purely absorbing, purely scattering media and those involving absorption and re-emission.

Wilson, Wan and Sen (1980) examined the possibility of extending Unno and Kondo's method by dividing the radiation field into three streams instead of two. The specific intensity representing each stream was an average over the appropriate zone. The division into three streams was inspired by the necessity to account for the shadowing effect of the central core star [cf. Chou and Tien (1963) and Leong and Sen (1971)] so that the exact boundary conditions could be used.

7.3 Moment Methods Based on Generalised Eddington Approximation

The radiative transfer in a non-grey, static spherical dust shell ($r_1 < r < R$) surrounding a core star of radius a ($a \leq r_1$) was studied. The angular moments J_ν, F_ν, K_ν, and L_ν and M_ν corresponding to $\mu^n (n = 0, 1, 2, 3, 4)$ were expressed in terms of three streams of radiation and from them the generalised Eddington approximation conditions were spelt out. The transfer equations were written in moment form and eliminating L_ν and M_ν from them by way of generalized Eddington relations, four equations in J_ν, F_ν, K_ν and $\mu_r(r, \nu)$ were obtained. As a first step, they were solved numerically to obtain J_ν, F_ν and K_ν. However, the direct numerical integration of equations is complicated by the fact that the slope of the μ_r variable near the lower boundary was very steep. To circumvent this difficulty, a two-shell approach was proposed (Wilson, Wan and Sen (1980)). In a small region near the lower boundary, the moment equations were solved under assigned μ_r. Beyond this region, the direct integration of four equations was done treating μ_r as an unknown dependent variable. At the interface between the two shells, continuity of J_ν, F_ν, K_ν and μ_r was assured. The method applied to the radiative transfer problems in a thin purely absorbing medium of finite thickness exhibited "peaking effect" of Eddington factor as expected. J_ν and K_ν/J_ν obtained by this method compared well with the values of the same quantities obtained by Unno and Kondo (1977, p. 699, model A1). Wilson and Sen (1980) extended the method to solve the problem of light scattering by an optically thin, inhomogeneous, spherically symmetric planetary atmosphere illuminated by solar radiation. The results were compared with those of Sobolev (1975) [cf. Sen and Wilson (1990) p. 146–167].

The generalised Eddington approximation method with three stream representation of radiation field has been used by Sen and Wilson (1993) for solving the moment equations of radiative transfer problems in a thin, non-grey, spherically symmetric, expanding atmosphere surrounding a core star. The method is outlined below for a simple atmospheric model.

In the transfer equation in a comoving frame we have ignored the terms involving aberration, advection and acceleration terms. Only the effects of Doppler shift has been considered. Under the circumstances, the radiative transfer equation in the spherically symmetric expanding envelope surrounding the core star can be written as [cf. equation (2.1.14) ; also Mihalas (1978) p. 492].

$$\mu[\partial I(r,\mu,\nu)/\partial r] + (1-\mu^2)/r[\partial I(r,\mu,\nu)/\partial \mu]$$
$$-(\nu V)/cr[(1-\mu^2) + \mu^2(d\ln V/d\ln r)]\partial I(r,\mu,\nu)/\partial \nu$$
$$= \eta(r,\nu) - \chi(r,\nu)I(r,\mu,\nu). \qquad (7.3.1)$$

In (7.3.1), all radiation and material quantities have been expressed in comoving frame. Here

$$\chi = k + \sigma, \quad \chi = \chi(r,\nu), \quad k = k(r,\nu), \quad \sigma = \sigma(r,\nu). \quad (7.3.2)$$

The source function,

$$S(r,\nu) = \eta/\chi \text{ with } \eta = \eta(r,\nu). \qquad (7.3.3)$$

k and σ are absorption and scattering coefficients respectively.

The source function $S(r,\nu)$ is given by

$$S(r,\nu) = \varepsilon B(r,\nu) + (1-\varepsilon)(\sigma/2)\int_{-1}^{1} I(r,\mu,\nu)d\mu. \qquad (7.3.4)$$

The symbols have their usual meanings. $\varepsilon = k/\chi$. The opacity χ and emissivity η are isotropic, that is angle independent in comoving frame. The following angle-averaged moments are introduced, all functions of (r,μ).

For example

$$[J,H,K,L,M,N] = \frac{1}{2}\int_{-1}^{1} d\mu I(r,\mu,\nu)\mu^n, \; [n=0,1,2,3,4,5]. \qquad (7.3.5)$$

From (7.3.1) and (7.3.5), we write the successive moment equations in comoving frame as

$$(1/r^2)[\partial(r^2 H)/\partial r] - \tilde{a}[\partial(J-K)/\partial \nu + b(\partial K/\partial \nu)] = \eta - \chi J \quad (7.3.6)$$

$$(\partial K/\partial r) + \frac{1}{r}(3K-J) - \tilde{a}[\partial(H-L)/\partial \nu + b(\partial L/\partial \nu)] = -\chi H \cdot \qquad (7.3.7)$$

$$(\partial L/\partial r) + \frac{2}{r}(2L-H) - \tilde{a}[\partial(K-M)/\partial \nu + b(\partial M/\partial \nu)] = \eta/3 - \chi K \qquad (7.3.8)$$

$$(\partial M/\partial r) + \frac{1}{r}(5M - 3K) - \tilde{a}[\partial(L - N)/\partial \nu + b(\partial N/\partial \nu)] = -\chi L \tag{7.3.9}$$

where

$$\tilde{a} = (\nu V)/(cr) \text{ and } b = (d\ln V/d\ln r). \tag{7.3.10}$$

The moment equation (7.3.6)–(7.3.9) suffer from their usual limitation of having more unknowns than the number of equations. Some closure conditions are needed to bridge this difference.

Closure Conditions and Generalised Eddington Relations

To obtain the closure conditions, we use the generalised Eddington approximation [cf. Unno and Kondo (1976); Wilson, Wan and Sen (1980)].

Three stream representation of the radiation field is made, each stream being the intensity average of the angles in the appropriate region. An angular parameter $\mu_r(r,\nu)$ is introduced in the process of division.

We represent $I(r,\mu,\nu)$ as

$$I(r,\mu,\nu) = \begin{cases} I_1(r,\nu) & \text{for } \mu_r < \mu \leq 1 \\ I_2(r,\nu) & \text{for } 0 < \mu \leq \mu_r \\ I_3(r,\nu) & \text{for } -1 \leq \mu \leq 0. \end{cases} \tag{7.3.11}$$

μ_r is unknown and is determined from the solution of moment equations. Then

$$J = \frac{1}{2}\int_{-1}^{1} I(r,\mu,\nu) d\mu = \frac{1}{2}[(I_1 + I_3) + \mu_r(I_2 - I_1)] \tag{7.3.12}$$

$$H = \frac{1}{2}\int_{-1}^{1} I(r,\mu,\nu)\mu \, d\mu = \frac{1}{4}[(I_1 - I_3) + \mu_r^2(I_2 - I_1)] \tag{7.3.13}$$

$$K = \frac{1}{2}\int_{-1}^{+1} I(r,\mu,\nu)\mu^2 d\mu = \frac{1}{6}[(I_1 + I_3) + \mu_r^3(I_2 - I_1)] \tag{7.3.14}$$

$$L = \frac{1}{2}\int_{-1}^{+1} I(r,\mu,\nu)\mu^3 d\mu = \frac{1}{8}[(I_1 - I_3) + \mu_r^4(I_2 - I_1)] \tag{7.3.15}$$

$$M = \frac{1}{2}\int_{-1}^{+1} I(r,\mu,\nu)\mu^4 d\mu = \frac{1}{10}[(I_1+I_3)+\mu_r^5(I_2-I_1)] \quad (7.3.16)$$

$$N = \frac{1}{2}\int_{-1}^{+1} I(r,\mu,\nu)\mu^5 d\mu = \frac{1}{12}[(I_1-I_3)+\mu_r^6(I_2-I_1)]. \quad (7.3.17)$$

From (7.3.12)–(7.3.17), we get the generalised Eddington relations.

$$2[2L - H] = \mu_r(3K - J) \quad (7.3.18)$$

$$(5M - 3K) = \mu_r^2(3K - J) \quad (7.3.19)$$

$$(6N - 4L) = \mu_r^3(3K - J). \quad (7.3.20)$$

These are the closure conditions for the moment equations (7.3.6)–(7.3.9) and they allow us to express L, M and N in terms of J, H, K and μ_r. Thus

$$L = \frac{1}{4}[\mu_r(3K - J) + 2H] \quad (7.3.21)$$

$$M = \frac{1}{5}[\mu_r^2(3K - J) + 3K] \quad (7.3.22)$$

$$N = \frac{1}{6}[\mu_r(\mu_r^2 + 1)(3K - J) + 2H]. \quad (7.3.23)$$

Now substituting (7.3.21)–(7.3.23) in (7.3.6)–(7.3.9) we have the following moment equations.

$$\frac{1}{r^2}[\partial(r^2 H)/\partial r] - \tilde{a}[\partial(J-K)/\partial\nu + b(\partial K/\partial\nu)] = \eta - \chi J \quad (7.3.24)$$

$$(\partial K/\partial r) + \frac{1}{r}(3K - J) - \frac{\tilde{a}}{2}[\partial\{H - (\mu_r/2)(3K - J)\}/\partial\nu$$
$$+ \frac{b}{2}\partial\{\mu_r(3K - J) + 2H\}/\partial\nu] = -\chi H \quad (7.3.25)$$

$$\frac{1}{4}[\partial\{\mu_r(3K - J) + 2H\}]/\partial r + \frac{1}{r}[\mu_r(3K - J)]$$
$$- \frac{\tilde{a}}{5}[\partial\{2K - \mu_r^2(3K - J)\}/\partial\nu + b\partial\{\mu_r^2(3K - J) + 3K\}/\partial\nu]$$
$$= (\eta/3) - \chi K \quad (7.3.26)$$

and

$$\frac{1}{5}\left[\partial\{\mu_r^2(3K-J)+3K\}/\partial r\right] + \frac{1}{r}[\mu_r^2(3K-J)]$$
$$-\frac{\tilde{a}}{12}\left[\partial\{(3K-J)\mu_r(1-2\mu_r^2)+2H\}/\partial\nu\right.$$
$$\left.+(2b)\partial\{(3K-J)\mu_r(\mu_r^2H)+2H\}/\partial\nu\right]$$
$$=-\frac{\chi}{4}\{\mu_r(3K-J)+2H\}. \qquad (7.3.27)$$

The set of four partial differential equations (7.3.24)–(7.3.27) now have four unknowns J, H, K, μ_r, which can be determined solving them under appropriate boundary conditions on r and initial conditions on ν. The velocity field law is supposed to be given.

Boundary Conditions and Initial Conditions

(a) The boundary conditions on radial coordinates

$$I(R,\mu,\nu) = \begin{cases} I_1(R,\nu) & \text{for } \mu_R < \mu \leq 1 \\ I_2(R,\nu) & \text{for } 0 < \mu \leq \mu_R \\ 0 & \text{for } -1 \leq \mu \leq 0 \end{cases} \qquad (7.3.28)$$

and

$$I(r_1,\mu,\nu) = \begin{cases} 0 & \text{for } \mu_{r_1} < \mu \leq 1 \\ I_2 = I(r_1,\nu) & \text{for } 0_1 < \mu \leq \mu_{r_1} \\ I_3 = I(r_1,\nu) & \text{for } -1 \leq \mu \leq 0. \end{cases} \qquad (7.3.29)$$

(b) Initial condition for ν. For the continuum I_1, I_2, I_3 are zero for

$$|\nu - \nu_0| \geq \nu_{\max}. \qquad (7.3.30)$$

(c) The velocity field in the medium is taken to be

$$V(r) = A(r - r_1), \quad V(R) = V_2. \qquad (7.3.31)$$

This implies the existence of velocity gradient.

Now from (7.3.12)–(7.3.17) and (7.3.28) and (7.3.29), we have

$$3K(R,\nu) - 2H(R,\nu)(1+\mu_R) + \mu_R J(R,\nu) = 0 \qquad (7.3.32)$$

$$3K(r_1,\nu)(1-\mu_{r_1}) + 2H(r_1,\nu)(1-\mu_{r_1}+\mu_{r_1}^2) = 0 \qquad (7.3.33)$$

$$3K(r_1,\nu) - J(r_1,\nu) - 2\mu_{r_1}H(r_1,\nu) = 0 \qquad (7.3.34)$$

with

$$\mu_{r_1} = \left(1 - \frac{a^2}{r_1^2}\right)^{1/2}. \qquad (7.3.35)$$

From the transfer equations (7.3.24)–(7.3.27), under the boundary conditions (7.3.32)–(7.3.34) and initial condition (7.3.30) and the knowledge of velocity field (7.3.31), we can find out J, H and K as a function of radius and frequency. Sen and Wilson (1993) numerically integrated the equations for an idealised model of thin shell atmosphere surrounding a purely absorbing core star. Tables of K/J against $z (z = r/a, 10, z < 60)$ is given for frequency $x = 2$ and $x = 4$, $V_2 = 0.01$ [cf. Sen and Wilson (1993) p. 234–240]. $J(R)$ is also plotted against frequency for $V_2 = 0.01$, $H(R)$ for $V_2 = 0.01, 0.05$ and 0.5. Variation of $H(R)$ with z for $V_2 = 0.01$ $x = 2$ and $x = 4$ has also been plotted. As the atmosphere is assumed thin, μ_r is taken to be

$$\mu_r = \left(1 - \frac{a^2}{r^2}\right)^{1/2}$$

throughout the shell.

For optically thick medium the two shell scheme viz a thin shell approximation for far interior and direct numerical integrations (7.3.24)–(7.3.27) for the outer region may be attempted. However no results are available so far.

References

Auer, L.M. (1971): JQSRT, **11**, 573.
Chandrasekhar, S. (1934): MNRAS, **94**, 443.
Chou, Y.S., Tien, C.L. (1963): JQSRT, **6**, 919.
Haish, B.M. (1979): Astronom. Astrophys., **72**, 161.
Hearn, A.G. (1964): Proc. Phys. Soc., **84**, 11.
Hummer, D.G. (1969): MNRAS, **145**, 95.

Leong, T.K., Sen, K.K. (1971): Publ. Astro. Soc. Japan, **23**, 99.
Magnan, C. (1974): Astronom. Astrophys., **35**, 233.
McCrea, W.H., Mitra, K.K. (1936): Zeitschr. Astro., f11, 359.
Mihalas, D., Kunasz, P.B., Hummer, D.G. (1976): Ap. J., **206**, 515.
Mihalas, D., Kunasz, P.B., Hummer, D.G. (1976): Ap. J., **210**, 419.
Mihalas, D., Kunasz, P.B. (1978): Ap. J., **219**, 635.
Mihalas, D. (1978): Stellar Atmospheres W. Freeman and Co., San Francisco.
Mihalas, D. (1980): Ap. J., **237**, 574.
Milkey, R.W., Mihalas, D. (1973): Ap. J., **185**, 709.
Sen, K.K., Wilson, S.J. (1990): Rad. Trans. in curved media, World Sci. Publ. Co., Singapore.
Sen, K.K., Wilson, S.J. (1993): Astrophy. Space Sc., **203**, 22.
Sobolev, V.V. (1975): Light Sca. in Planetary Atms., Pergamon Press. Oxford.
Unno, W., Kondo, M. (1976): Publ. Astro. Soc. Japan, **28**, 347.
Unno, W., Kondo, M. (1977): Publ. Astro. Soc. Japan, **29**, 673.
Wilson, S.J., Wan, F.S., Sen K.K. (1980): Astrophy. Space Sc., **67**, 99.
Wilson, S.J., Sen, K.K. (1980): Astrophy. Space Sc., **71**, 405.

Chapter 8. Numerical Methods for Transfer Equations in CMF

8.1 Integral Operator Technique

Discrete ordinate method developed by Chandrasekhar (1960) and Kourganoff (1963) has proved to be a very efficient way to solve plane parallel transfer problems. There has been several attempts to extend this idea of discretizing the transfer equation for transfer problems in spherical media. Peraiah and Grant (1973) attempted to use the discrete space theory, proposed by Grant and Hunt (1968) for the spherical shell media. This cell method, as it is called, gave very good results provided the dimensions of the radius-angle grid is small. We shall now describe the main ideas of this method as detailed by Peraiah in his article in *Methods in Radiative Transfer*, edited by W. Kalkofen (1984). The spherical medium under consideration is first divided into shells by the radial coordinate r and let $U_i = (r_i^2 I_i)$ denote the intensity like variable at r_i, $i = 1, 2, \ldots D$. The emergent and incident intensities at the surfaces of the elementary shells are expressed in terms of reflection and transmission operators in a way resembling that in Ambarzumian's physical method. An "interaction principle" is introduced through "S-matrices", which are connected by "star products". Internal radiation field is calculated in terms of the reflection and transmission operators between the boundaries of the successive cells. In turn these operators are determined by integral operator techniques. For the present we suppress the dependence of the intensity on the angle and frequency.

In the cell (r_i, r_{i+1}), the intensity in the positive outward direction ($\mu > 0$) is denoted by U^+ and the intensity in the opposite inward direction ($\mu < 0$) by U^-. Thus the incident intensities are U_i^- and U_{i+1}^+ and the emergent intensities are U_i^+ and U_{i+1}^-.

196   8. Numerical Methods for Transfer Equations in CMF

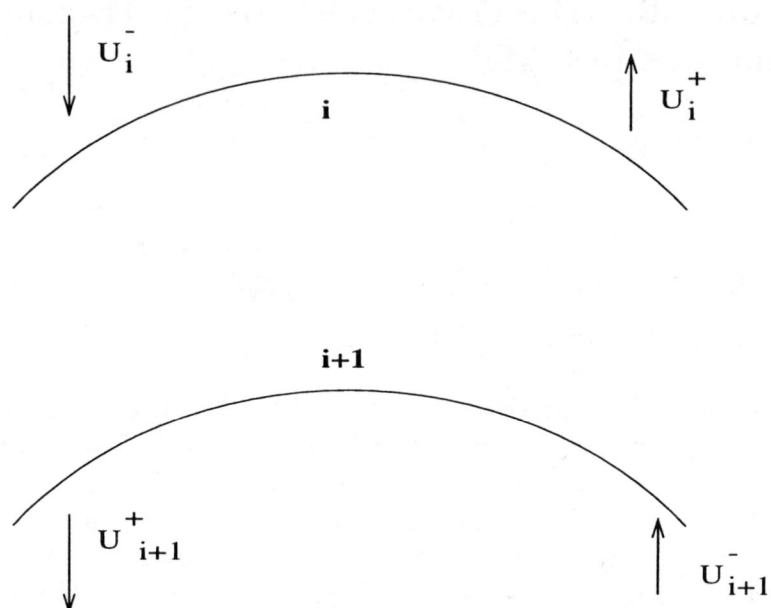

Figure 8.1. Intensity ray entering into and emerging from a shell envelope

These emergent and incident intensities are linked by the transmission and reflection operators $\mathbf{t}, \mathbf{r}$ and are given by

$$U_{i+1}^- = \mathbf{t}(i+1,i)U_i^- + \mathbf{r}(i,i+1)U_{i+1}^+ + \Sigma^-(i+1,i)$$
$$U_i^+ = \mathbf{r}(i+1,i)U_i^- + \mathbf{t}(i,i+1)U_{i+1}^+ + \Sigma^+(i,i+1). \quad (8.1.1)$$

Here $\Sigma^-(i+1,i)$ and $\Sigma^+(i,i+1)$ are the contributions from the truly internal sources within this cell. The first index denotes the surface where the intensity is calculated and this together with the second index describes the cell considered. The above equation (8.1.1) is called the "Interaction Principle". This can be rewritten as

8.1 Integral Operator Technique

$$\begin{pmatrix} U_{i+1}^- \\ U_i^+ \end{pmatrix} = S(i, i+1) \begin{pmatrix} U_i^- \\ U_{i+1}^+ \end{pmatrix} + \begin{pmatrix} \Sigma^-(i+1, i) \\ \Sigma^+(i, i+1) \end{pmatrix} \quad (8.1.2a)$$

where

$$S(i, i+1) = \begin{pmatrix} \mathbf{t}(i+1, i) & \mathbf{r}(i, i+1) \\ \mathbf{r}(i+1, i) & \mathbf{t}(i, i+1) \end{pmatrix}. \quad (8.1.2b)$$

If there is another adjacent cell with boundaries $(i+1, i+2)$, then we have by the Interaction Principle

$$\begin{pmatrix} U_{i+2}^- \\ U_{i+1}^+ \end{pmatrix} = S(i+1, i+2) \begin{pmatrix} U_{i+1}^- \\ U_{i+2}^+ \end{pmatrix} + \begin{pmatrix} \Sigma^-(i+2, i+1) \\ \Sigma^+(i+1, i+2) \end{pmatrix}.$$

$$(8.1.3)$$

If we now combine the two cells, then again from the Interaction Principle, we must have

$$\begin{pmatrix} U_{i+2}^- \\ U_i^+ \end{pmatrix} = S(i, i+2) \begin{pmatrix} U_i^- \\ U_{i+2}^+ \end{pmatrix} + \begin{pmatrix} \Sigma^-(i+2, i) \\ \Sigma^+(i, i+2) \end{pmatrix}. \quad (8.1.4)$$

But equation (8.1.4) can be also obtained from (8.1.2) and (8.1.3) by eliminating U_{i+1}^+, and U_{i+1}^-. By this process the S-matrices will be linked and this relationship can be written in the form

$$S(i, i+2) = S(i, i+1) * S(i+1, i+2),$$

and is called the "star product" of the S-matrices. As a consequence we obtain the relationship between the transmission and reflection operators as

$$\begin{aligned}
\mathbf{t}(i+2, i) &= \mathbf{t}(i+2, i+1)[\mathbb{I} - \mathbf{r}(i, i+1)\mathbf{r}(i+2, i+1)]^{-1} \\
&\quad \times \mathbf{t}(i+1, i) \\
\mathbf{t}(i, i+2) &= \mathbf{t}(i, i+1)[\mathbb{I} - \mathbf{r}(i+2, i+1)\mathbf{r}(i, i+1)]^{-1} \\
&\quad \times \mathbf{t}(i+1, i+2) \\
\mathbf{r}(i+2, i) &= \mathbf{r}(i+1, i) + \mathbf{t}(i, i+1)\mathbf{r}(i+2, i+1) \\
&\quad \times [\mathbb{I} - \mathbf{r}(i, i+1)\mathbf{r}(i+2, i+1)]^{-1} \mathbf{t}(i+1, i) \\
\mathbf{r}(i, i+2) &= \mathbf{r}(i+1, i+2) + \mathbf{t}(i+2, i+1)\mathbf{r}(i, i+1) \\
&\quad \times [\mathbb{I} - \mathbf{r}(i+2, i+1)\mathbf{r}(i, i+1)]^{-1} \mathbf{t}(i+1, i+2);
\end{aligned}$$
$$(8.1.5)$$

where $\mathbb{I}$ is the identity matrix.

In particular, by considering the adjacent cells as $(1,i)$ and $(i,i+1)$, we have from (8.1.5)

$$\begin{aligned}\mathbf{r}(1,i+1) &= \mathbf{r}(i,i+1) + \mathbf{t}(i+1,i)\mathbf{r}(1,i) \\ &\quad \times [\mathbb{I} - \mathbf{r}(i+1,i)\mathbf{r}(1,i)]^{-1}\,\mathbf{t}(i,i+1).\end{aligned} \qquad (8.1.6)$$

We shall now describe Grant and Hunt's (1968) scheme of calculating the internal radiation field in terms of the known transmission and reflection operators between the boundaries of the cell $(i,i+1)$. We will show later how these transmission and reflection operators $\mathbf{t}(i,i+1)$, $\mathbf{t}(i+1,i)$, $\mathbf{r}(i,i+1)$, $\mathbf{r}(i+1,i)$ are computed using the integral operator technique. The internal intensities are computed with the formula for $i = D, D-1, D-2, \ldots, 2, 1$ given by

$$\begin{aligned} U_{i+1}^- &= \mathbf{r}(1,i+1)U_{i+1}^+ + V_{i+1/2}^- \\ U_i^+ &= \hat{\mathbf{t}}(i,i+1)U_{i+1}^+ + V_{i+1/2}^+ \end{aligned}$$

with the initial condition

$$U_{D+1}^+ = U^+(a) \qquad (8.1.7)$$

where $r = a$ is the inner spherical boundary of the medium and $\hat{\mathbf{t}}$ will be defined subsequently. The operators on the right side of equation (8.1.7) are computed sequentially as follows. The following operators are first defined.

$$\begin{aligned} \mathbb{T}_{i+1/2} &= [\mathbb{I} - \mathbf{r}(i+1,i)\mathbf{r}(1,i)]^{-1} \\ \mathbb{K}_{i+1/2} &= [\mathbb{I} - \mathbf{r}(1,i)\mathbf{r}(i+1,i)]^{-1} \\ \hat{\mathbf{t}}(i+1,i) &= \mathbf{t}(i+1,i)\mathbb{K}_{i+1/2} \\ \hat{\mathbf{r}}(i+1,i) &= \mathbf{r}(i+1,i)\mathbb{K}_{i+1/2} \\ \mathbb{R}_{i+1/2} &= \hat{\mathbf{t}}(i+1,i)\mathbf{r}(1,i). \end{aligned} \qquad (8.1.8)$$

Then the following operators are computed sequentially for $i = 1, 2, \ldots, D$

$$\mathbf{r}(1,i+1) = \mathbf{r}(i,i+1) + \mathbf{t}(i+1,i)\mathbf{r}(1,i)\mathbb{T}_{i+1/2}\mathbf{t}(i,i+1) \qquad (8.1.9a)$$

$$V^-_{i+1/2} = \hat{\mathbf{t}}(i+1,i)V^-_{i-1/2} + \Sigma^-(i+1,i) + \mathbb{R}_{i+1/2}\Sigma^+(i,i+1) \quad (8.1.9b)$$

$$V^+_{i+1/2} = \hat{\mathbf{r}}(i+1,i)V^-_{i+1/2} + \mathbb{T}_{i+1/2}\Sigma^+(i,i+1) \quad (8.1.9c)$$

with initial condition

$$\mathbf{r}(1,1) = 0$$

and

$$V^-_{1/2} = U^-(b)$$

where $r = b$ is the outer boundary of the medium. We note that (8.1.9a) is in fact equation (8.1.6) obtained earlier. The operators V denote the cumulative contributions of the internal sources. This is clear from equation (8.1.7) and they can be computed using the transmission and reflection operators and are given precisely by equations (8.1.9b) and (8.1.9c). This completes Grant and Hunt's scheme of calculating the internal radiation. They noted that unless the mesh was very small positive and negative elements of the matrix obtained by the discretization of the curvature term gave rise to unphysical negative intensities in the final solution. To overcome this limitation Peraiah and Varghese (1985) proposed the integral operator technique which is again in the flavour of the discrete ordinate scheme but introduces integrations over the mesh by suitable operators before the numerical solution is sought. Our main reference for this review will be from Peraiah and Varghese (1985), Peraiah (1990,1991) and Peraiah's article in Numerical Radiative Transfer [cf. Kalkofen (1987)]. We shall now outline this method for the spherically symmetric expanding envelope where the expansion speed $\beta = v/c$ is assumed to be small. (In this section we have used the letter v instead of V to indicate the speed). In the comoving frame, the time independent transfer equation can be written as [cf. (2.5.19)]

$$\frac{v}{c}\frac{\partial I}{\partial r} + \frac{\mu}{r^2}\frac{\partial}{\partial r}(r^2 I) + \frac{\partial}{\partial \mu}\left[(1-\mu^2)\left\{\frac{1}{r} + \frac{\mu}{c}(\frac{v}{r} - \frac{\partial v}{\partial r})\right\}I\right]$$

$$-\frac{\partial}{\partial \nu}\left[\nu\left\{(1-\mu^2)\frac{v}{cr} + \frac{\mu^2}{c}\frac{\partial v}{\partial r}\right\}I\right]$$

$$+\left[(3-\mu^2)\frac{v}{cr} + \frac{1+\mu^2}{c}\frac{\partial v}{\partial r}\right]I$$

$$= \eta - \chi I. \quad (8.1.10)$$

8. Numerical Methods for Transfer Equations in CMF

Writing $U = r^2 I$, $S = \eta/\chi$ and $\beta = v/c$, the above equation can be written as

$$(\beta + \mu)\frac{\partial U(r,\mu,\nu)}{\partial r} + \frac{1-\mu^2}{r}\left[1 + \mu\beta\left\{1 - \frac{r}{\beta}\frac{d\beta}{dr}\right\}\frac{\partial U}{\partial \mu}(r,\mu,\nu)\right]$$
$$- \left[\nu\left\{(1-\mu^2)\frac{\beta}{r} + \mu^2\frac{d\beta}{dr}\right\}\right]\frac{\partial U}{\partial \nu}(r,\mu,\nu)$$
$$+ \left\{3\left[\frac{\beta}{r}(1-\mu^2) + \mu^2\frac{d\beta}{dr}\right] - \frac{2(\mu+\beta)}{r}\right\}U(r,\mu,\nu)$$
$$= \chi[S(r,\mu,\nu) - U(r,\mu,\nu)]. \qquad (8.1.11)$$

In the above equation $\mu \in (0,1)$. We may in the same way write down a similar equation for the opposite directed ray, still keeping μ in the range $(0,1)$.

We may introduce the radius-angle-frequency grid r_i, μ_j, ν_k and consider the elemental volume shown below.

Next, we introduce an interpolating polynomial within the mesh. We express the intensity U within this volume by the interpolating multinomial formula in the variables ξ, η and ζ as follows:

$$U(r,\mu,\nu) = U_0 + U_r\xi + U_\mu\eta + U_\nu\zeta + U_{r\mu}\xi\eta + U_{\mu\nu}\eta\zeta$$
$$+ U_{\nu r}\zeta\xi + U_{r\mu\nu}\xi\eta\zeta, \qquad (8.1.12a)$$

where

$$\xi = \frac{r-\bar{r}}{\Delta r/2}, \qquad \eta = \frac{\mu-\bar{\mu}}{\Delta\mu/2}, \qquad \zeta = \frac{\nu-\bar{\nu}}{\Delta\nu/2}$$
$$\bar{r} = \tfrac{1}{2}(r_i + r_{i-1}), \qquad \Delta r = (r_i - r_{i-1}),$$
$$\bar{\mu} = \tfrac{1}{2}(\mu_j + \mu_{j-1}), \qquad \Delta\mu = (\mu_j - \mu_{j-1}),$$
$$\bar{\nu} = \tfrac{1}{2}(\nu_k + \nu_{k-1}), \qquad \Delta\nu = (\nu_k - \nu_{k-1}).$$

$$(8.1.12b)$$

Introducing the vector **A** to describe the eight interpolating constant coefficients defined above and the vector **B** to describe the eight nodal values of the intensities at the vertices of the elemental volume, we have

$$\mathbf{A} = (U_0, U_r, U_\mu, U_\nu, U_{r\mu}, U_{\mu\nu}, U_{\nu r}, U_{r\mu\nu})^T$$
$$\mathbf{B} = (U_a, U_b, U_c, U_d, U_e, U_f, U_g, U_h)^T.$$

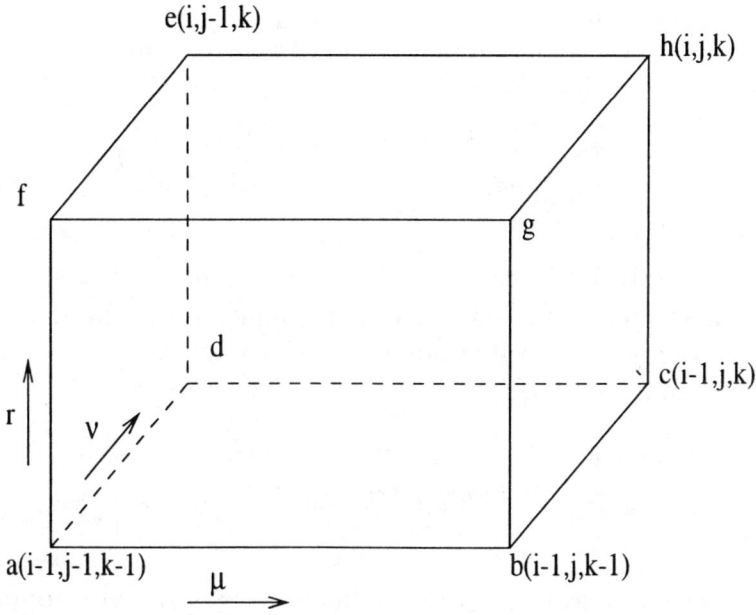

Figure 8.2. Schematic diagram of the angle-frequency-radius grid

By using the relations in (8.1.12) it can be easily seen that these vectors are related by the equation

$$\mathbf{A} = \frac{1}{8}\mathbb{T}\mathbf{B} \tag{8.1.13}$$

where $\mathbb{T}$ is the matrix given by

$$\mathbb{T} = \begin{pmatrix} 1 & 1 & 1 & 1 & 1 & 1 & 1 & 1 \\ -1 & -1 & -1 & -1 & 1 & 1 & 1 & 1 \\ -1 & 1 & 1 & -1 & -1 & -1 & 1 & 1 \\ -1 & -1 & 1 & 1 & 1 & -1 & -1 & 1 \\ 1 & -1 & -1 & 1 & -1 & -1 & 1 & 1 \\ 1 & -1 & 1 & -1 & -1 & 1 & 1 & 1 \\ 1 & 1 & -1 & -1 & 1 & -1 & -1 & 1 \\ -1 & 1 & -1 & 1 & -1 & 1 & -1 & 1 \end{pmatrix} \tag{8.1.14}$$

Equation (8.1.13) expresses the interpolating coefficients in terms of the nodal values of the intensity.

The final step and the main idea of the present scheme is the introduction of the integral operators X,Y and Z given by

$$\begin{array}{ll} X = \frac{1}{V}\int_{\triangle r}(\cdot)4\pi r^2 dr, & V = \frac{4\pi}{3}(r_i^3 - r_{i-1}^3) \\ Y = \frac{1}{\triangle \mu}\int_{\triangle \mu}(\cdot)d\mu, & Z = \frac{1}{\triangle \nu}\int_{\triangle \nu}(\cdot)d\nu. \end{array} \quad (8.1.15)$$

In the numerical scheme proposed by Peraiah, the transfer equation is first approximated by the interpolating multinomials over the mesh. The source function S is then approximated by interpolating multinomials in the variables ξ, η and ζ as

$$\begin{aligned} S(r,\mu,\nu) = & S_0 + \xi S_r + \eta S_\mu + \zeta S_\nu \\ & + \xi\eta S_{r\mu} + \eta\zeta S_{\mu\nu} + \zeta\xi S_{\nu r} + \xi\eta\zeta S_{r\mu\nu}; \end{aligned} \quad (8.1.16)$$

Finally the integral operators defined in (8.1.15) are applied in succession to the approximate transfer equation. In going through these steps, the opacity is assumed to be constant over the elemental radial distance $(r_i - r_{i-1})$.

The interpolating coefficients of the intensity $U(r,\mu,\nu)$ occurring in the vector $\mathbf{A}$ can be expressed in terms of the nodal values occurring in vector $\mathbf{B}$ using the relation (8.1.13). Further as the contribution to the source term $S(r,\mu,\nu)$ arises from both the scattering integral and the truly generated internal sources, the interpolating coefficients of the source function $S(r,\mu,\nu)$ can be also expressed in terms of the nodal values and the truly generated internal sources. Let us now write the nodal values in their full form such as

$$U_a^+ = U_{i-1,j-1,k-1}^+, \quad U_b^+ = U_{i-1,j,k-1}^+, \quad \text{etc.} \quad (8.1.17)$$

and denote the integrated truly internal source flowing from the cell in the outward direction as $\Sigma^+(i-1, j-1, k-1, i, j, k)$. Introducing the column vector $\mathbf{U}_i^+$ whose elements are $U_{i,j,k}^+$; $j = 1,\ldots, J$; $k = 1, 2, \ldots K$, we can rewrite radius, angle, frequency integrated approximate equation [via (8.1.15)] in the form

$$\mathbb{A}_1 \mathbf{U}_{i+1}^+ + \mathbb{B}_1 \mathbf{U}_i^+ = \mathbb{C}_1 \mathbf{U}_{i+1}^- + \mathbb{D}_1 \mathbf{U}_i^- + \mathbb{E}_1 \Sigma^+ \quad (8.1.18)$$

where $\mathbb{A}_1, \mathbb{B}_1, \mathbb{C}_1, \mathbb{D}_1, \mathbb{E}_1$ are known matrices.

Similarly starting from the transfer equation for the inward flowing radiation one can obtain a similar equation in the form

$$\mathbb{A}_2 \mathbf{U}^-_{i+1} + \mathbb{B}_2 \mathbf{U}^-_i = \mathbb{C}_2 \mathbf{U}^+_{i+1} + \mathbb{D}_2 \mathbf{U}^+_i + \mathbb{E}_2 \mathbf{\Sigma}^-. \qquad (8.1.19)$$

The above two equations can be put in the form

$$\begin{pmatrix} \mathbf{U}^-_{i+1} \\ \mathbf{U}^+_i \end{pmatrix} = \begin{pmatrix} X_1 & X_2 \\ X_3 & X_4 \end{pmatrix} \begin{pmatrix} \mathbf{U}^-_i \\ \mathbf{U}^+_{i+1} \end{pmatrix} + \begin{pmatrix} \mathbf{Z}_1 \\ \mathbf{Z}_2 \end{pmatrix} \qquad (8.1.20)$$

where $\mathbf{U}^-_{i+1}, \mathbf{U}^+_i$ are the output intensities; $\mathbf{U}^-_i, \mathbf{U}^+_{i+1}$ are the incident intensities and $\mathbf{Z}_1, \mathbf{Z}_2$ are the contributions arising from the internal sources.

Comparing equation (8.1.20) with the equation (8.1.2) obtained from the interaction principle, the transmission and reflection operators $\mathbf{t}(i+1,i), \mathbf{t}(i,i+1), \mathbf{r}(i,i+1), \mathbf{r}(i+i,i)$ can be obtained. Details of these can be found in Peraiah and Varghese (1985) and Peraiah (1990). Once these operators are known the internal intensities can be computed by the scheme given by equations (8.1.7), (8.1.8) and (8.1.9). This completes the integral operator technique to solve for the intensities in a spherically symmetric expanding envelope.

8.2 Adaptive Mesh Method

In numerical simulation of fluid flows and radiation hydrodynamics, Winkler *et al.* (1983, 1985), Mihalas *et al.* (1984a, 1984b, 1984c) have shown the advantages of using an adaptive mesh in which the coordinate system is fixed neither in the laboratory inertial frame nor in the comoving fluid frame. This scheme of allowing the mesh to evolve freely in such a manner as to track significant physical features of the flow led to efficient and accurate computational schemes. In the context of spherical atmospheres, the evolution of the mesh is governed by two distinct terms, one called the radial function f^r which ensures the global consistency of the mesh and the other called the structure function f^s which provides a measure of the physical structure of the flow problem at hand. The main difficulty in using the adaptive mesh technique is in the formulation of the problem in a conservative form for the adaptive coordinate system. In the following we will first discuss the radial function f^r and the structure function f^s occurring in the evolution of the mesh and then obtain the transfer equations in the conservative form to be used with the adaptive mesh. Mihalas' work with Winkler and Norman (1983, 1984a, 1984b, 1984c, 1985) are the main references for this review. We will consider a spherically expanding atmosphere of finite thickness and let r_i be the radius of a typical grid. The equation governing the grid motion is written in the form

$$(\Delta f^r)_i + (\Delta f^s)_i - \sigma_g \left(\frac{\delta r_i}{\delta t}\right) = 0, \qquad (8.2.1)$$

where f^r is the radial function f^s is the structure function and σ_g a positive constant. These terms will be defined subsequently. Here the spatial difference Δ is zone centered, i.e. $(\Delta f)_i = f_{i+1} - f_i$.

The radial function f_i^r has essentially four terms and is given by

$$f_i^r = W_r \Delta r_i + W_{rz} \frac{(\Delta r_i)^2}{(\Delta r_{i-1} \Delta r_{i+1})}$$
$$+ W_{r_{max}} \left(\frac{\Delta r_i}{\Delta r_{max}}\right)^{\bar{n}} - W_{r_{min}} \left(\frac{\Delta r_{min}}{\Delta r_i}\right)^{\bar{n}}, \qquad (8.2.2)$$

where $\Delta r_i = (r_i - r_{i-1})/R_{scale}$.

The $W's$ represent the weights assigned to each term and the parameters $R_{scale}, \Delta r_{max}, \Delta r_{min}$ respectively set the scale of the given problem and the maximum, minimum zone sizes. The free parameter $\bar{n}$ is typically 4. The first term in (8.2.2) defines the spatial grid while the second term in (8.2.2) stablises the mesh and guarantees the monotonicity of the grid spacing. The other desirable properties of this particular stabilising term is discussed in detail by Winkler, Mihalas and Norman (1985). The constant σ_g occurring in the diffuse term $\sigma_g \frac{\delta r}{\delta t}$ allows the grid to evolve slowly except when σ_g is set to zero in which case the change is instantaneous. This time-filtering constant is defined as follows:

$$\sigma_g = \text{CVMGP } (0., 1., \delta f) \qquad (8.2.3a)$$

where
$$\delta f = f^s - f^n, \qquad (8.2.3b)$$
and
$$\text{CVMGP } (x, y, z) = \begin{cases} x & \text{if } z > 0 \\ y & \text{if } z \leq 0. \end{cases} \qquad (8.2.3c)$$

f^n is the reference value which is revised with each time step n to

$$f^{n+1} = f^s + (1-\varepsilon)(f^n - f^s) \text{ CVMGP } (0., 1., \delta f). \qquad (8.2.3d)$$

The constant ε plays the role of retaining the memory of the structure of the flow a few time steps later.

The structure function f^s occurring in the evolution equation (8.2.1) is designed to resolve all the important flow variables equally well in all the different zones of the mesh. This is done by taking the solution at the new time level n as the starting point for the adaptive mesh equation and making the solution, in the average sense, vary by the same factor from grid point to grid point everywhere in the domain. The following is the structure function suggested by Winkler et al. (1983). They have

$$\begin{aligned} f^s_i &= W_m \Delta m_i + W_{m\ell} \Delta m\ell_i + W_{\rho\ell} \Delta \rho\ell_i \\ &\quad + W_{P\ell} \Delta P\ell_i + W_{k\rho\ell} \Delta k\rho\ell_i \\ &\quad + W_{E\ell} \Delta E\ell_i + W_v \Delta \bar{V}_i + W_{v\ell} \Delta \bar{V}\ell_i \end{aligned} \qquad (8.2.4a)$$

where
$$\begin{aligned} \Delta m_i &= (m_i - m_{i+1})/M_{scale} \\ \Delta m\ell_i &= (m_i - m_{i+1})/(m_i + m_{i+1}) \\ \Delta \rho\ell_i &= [(\rho_{i-1} - \rho_{i+1})/\rho_i]^2 \\ \Delta P\ell_i &= [(P_{i-1} - P_{i+1})/P_i]^2 \\ \Delta k\rho\ell_i &= [(k_{i-1}\rho_{i-1} - k_{i+1}\rho_{i+1})/k_i\rho_i]^2 \\ \Delta E\ell_i &= [(E_{i-1} - E_{i+1})/E_i]^2 \\ \Delta \bar{V}_i &= (\bar{V}_i - \bar{V}_{i+1})^2 \text{ CVMGP } (1, 0, \bar{V}_i - \bar{V}_{i+1}) \\ \Delta \bar{V}\ell_i &= \frac{(\bar{V}_i - \bar{V}_{i+1})^2}{[\bar{V}_i^2 + \bar{V}_{i+1}^2 + c\bar{V}\ell]} \text{ CVMGP } (1, 0, \bar{V}_i - \bar{V}_{i+1}). \end{aligned}$$
$$(8.2.4b)$$

Here the symbol $X\ell = \ln X$ for any of the variables X and $\bar{V} = V_{rel}$.

Let us next obtain the adaptive mesh radiative transfer equations. These have been derived by Mihalas et al. (1984b). We detail them in the following paragraphs.

We have used two different time derivatives namely the Eulerian derivative ($\frac{\partial}{\partial t}$) and the Lagrangian derivative ($\frac{D}{Dt}$). If $\mathbf{V}$ is the fluid velocity then we know that for any differentiable function f these two derivatives are linked by the relation

$$\frac{Df}{Dt} = \frac{\partial f}{\partial t} + (\mathbf{V} \cdot \boldsymbol{\nabla})f. \qquad (8.2.5a)$$

We shall now denote the adaptive-mesh derivative taken with respect to fixed values of the moving adaptive-mesh coordinates by ($\frac{d}{dt}$). Then the grid velocity is given by

$$\mathbf{V}_g = \frac{dr}{dt};$$

and we obtain, as in (8.2.5a), the relation

$$\left(\frac{df}{dt}\right) = \left(\frac{\partial f}{\partial t}\right) + (\mathbf{V}_g \cdot \boldsymbol{\nabla})f. \qquad (8.2.5b)$$

Let us now define the relative velocity $\mathbf{V}_{rel}$ of the fluid with respect to the adaptive-mesh grid by

$$\mathbf{V}_{rel} = \mathbf{V} - \mathbf{V}_g. \qquad (8.2.6)$$

In fluid flow (cf. Aris 1962), using (8.2.5a) one obtains the Reynolds transport theorem in the form

$$\frac{D}{Dt}\left(\int_{V \text{ fluid}} f \, dV_{\text{fluid}}\right) = \int_{V \text{ fluid}} \left[\frac{\partial f}{\partial t} + \boldsymbol{\nabla} \cdot (\mathbf{V}f)\right] dV_{\text{fluid}}. \qquad (8.2.7)$$

In a similar way, using (8.2.5) one can easily obtain the relation

$$\int_V \rho \frac{D}{Dt}\left(\frac{f}{\rho}\right) dV = \frac{d}{dt} \int_V f \, dV + \int_{\partial V} f \mathbf{V}_{rel} \cdot d\mathbf{S}; \qquad (8.2.8)$$

8.2 Adaptive Mesh Method

where V is the adaptive-mesh volume enclosed by the surface ∂V which has an outward pointing surface element $\mathbf{dS}$. The above results will be used to write down the radiative transfer equation in the adaptive mesh coordinates. We have from equation (2.5.19), the radiative transfer equation for an expanding spherical atmosphere as

$$\frac{1}{c}\left[\frac{\partial I}{\partial t} + v\frac{\partial I}{\partial r}\right] + \frac{\mu}{r^2}\frac{\partial}{\partial r}(r^2 I)$$

$$+ \frac{\partial}{\partial \mu}\left[(1-\mu^2)\left\{\frac{1}{r} + \frac{\mu}{c}(\frac{v}{r} - \frac{\partial v}{\partial r}) - \frac{a}{c^2}\right\}I\right]$$

$$- \frac{\partial}{\partial \nu}\left[\nu\left\{(1-\mu^2)\frac{v}{cr} + \frac{\mu^2}{c}\frac{\partial v}{\partial r} + \frac{\mu a}{c^2}\right\}I\right]$$

$$+ \left[(3-\mu^2)\frac{v}{cr} + (\frac{1+\mu^2}{c})\frac{\partial v}{\partial r} + \frac{2\mu a}{c^2}\right]I - \eta + \chi I = 0.$$

(8.2.9)

Introducing the density ρ and setting $d\text{Vol} = \frac{1}{3}d(r^3)$, equation (8.2.9) can be written in terms of the Lagrangian derivative $\frac{D}{Dt}$ as

$$\left(\frac{\rho}{c}\right)\frac{D}{Dt}\left(\frac{I}{\rho}\right) + \frac{\partial}{\partial \text{Vol}}(\mu r^2 I)$$

$$+ \frac{\partial}{\partial \mu}\left[(1-\mu^2)\left\{\frac{1}{r} + \frac{\mu}{c}\left(\frac{3v}{r} - \frac{\partial(r^2 v)}{\partial \text{Vol}}\right) - \frac{a}{c^2}\right\}I\right]$$

$$- \frac{\partial}{\partial \nu}\left[\nu\left\{(1-3\mu^2)\frac{v}{cr} + \frac{\mu^2}{c}\frac{\partial(r^2 v)}{\partial \text{Vol}} + \frac{\mu a}{c^2}\right\}I\right] - \eta$$

$$+ \left[\chi + \left\{(1-3\mu^2)\frac{v}{cr} + \frac{\mu^2}{c}\frac{\partial(r^2 v)}{\partial \text{Vol}} + \frac{2\mu a}{c^2}\right\}\right]I = 0.$$

(8.2.10)

Let us now set

$$\tilde{\mathbf{V}} = (v, 0, 0) \quad \text{where} \quad v = \frac{Dr}{Dt}$$

$$\tilde{\mathbf{V}}_g = \left(V_g, \frac{d\mu}{dt}, \frac{d\nu}{dt}\right) \quad \text{where} \quad V_g = \frac{dr}{dt}$$

8. Numerical Methods for Transfer Equations in CMF

$$\tilde{\mathbf{V}}_{\text{rel}} = \tilde{\mathbf{V}} - \tilde{\mathbf{V}}_g$$
$$= \left(v - \frac{dr}{dt}, -\frac{d\mu}{dt}, -\frac{d\nu}{dt}\right)$$

and

$$\tilde{\boldsymbol{\nabla}} = \left(\frac{\partial}{\partial r}, \frac{\partial}{\partial \mu}, \frac{\partial}{\partial \nu}\right).$$

Then using equation (8.2.5) the equation (8.2.10) can be written in the adaptive mesh coordinates as

$$\frac{1}{c}\frac{dI}{dt} + (\tilde{\mathbf{V}}_{\text{rel}} \cdot \tilde{\boldsymbol{\nabla}})I + I(\boldsymbol{\nabla} \cdot \mathbf{V}) + \frac{\partial}{\partial \text{Vol}}(\mu r^2 I)$$
$$+ \frac{\partial}{\partial \mu}\left[(1-\mu^2)\left\{\frac{1}{r} + \frac{\mu}{c}\left(\frac{3v}{r} - \frac{\partial(r^2 v)}{\partial \text{Vol}}\right) - \frac{a}{c^2}\right\}I\right]$$
$$- \frac{\partial}{\partial \nu}\left[\nu\left\{(1-3\mu^2)\frac{v}{cr} + \frac{\mu^2}{c}\frac{\partial(r^2 v)}{\partial \text{Vol}} + \frac{\mu a}{c^2}\right\}I\right] - \eta$$
$$+ \left[\chi + \left\{(1-3\mu^2)\frac{v}{cr} + \frac{\mu^2}{c}\frac{\partial(r^2 v)}{\partial \text{Vol}} + \frac{2\mu a}{c^2}\right\}\right]I = 0.$$
(8.2.11)

The above equations reduce to the comoving frame equation (8.2.10) when $\tilde{\mathbf{V}}_{\text{rel}} = 0$. Introducing the adaptive-mesh volume $d\tilde{V} = d\text{Vol}d\mu d\nu$, with the corresponding surface element $\tilde{\mathbf{S}}$ and using the relation (8.2.8), equation (8.2.11) can be integrated over the mesh volume $d\tilde{V}$ to obtain the adaptive-mesh transfer equation in the conservative form as

$$\frac{d}{dt}\left[\int_{\tilde{V}}(\frac{I}{c})d\tilde{V}\right] + \int_{\partial\tilde{V}}(\frac{I}{c})(\tilde{\mathbf{V}}_{\text{rel}} \cdot d\tilde{\mathbf{S}}) + \int_{\tilde{V}}\frac{\partial}{\partial \text{Vol}}(\mu r^2 I)d\tilde{V}$$
$$+ \int_{\tilde{V}}\frac{\partial}{\partial \mu}\left[(1-\mu^2)\left\{\frac{1}{r} + \frac{\mu}{c}\left(\frac{3v}{r} - \frac{\partial(r^2 v)}{\partial \text{Vol}}\right) - \frac{a}{c^2}\right\}I\right]d\tilde{V}$$
$$- \int_{\tilde{V}}\frac{\partial}{\partial \nu}\left[\nu\left\{(1-3\mu^2)\frac{v}{cr} + \frac{\mu^2}{c}\frac{\partial(r^2 v)}{\partial \text{Vol}} + \frac{\mu a}{c^2}\right\}I\right]d\tilde{V}$$
$$+ \int_{\tilde{V}}\left[-\eta + \left\{\chi + (1-3\mu^2)\frac{v}{cr}\right.\right.$$
$$\left.\left. + \frac{\mu^2}{c}\frac{\partial(r^2 v)}{\partial \text{Vol}} + \frac{2\mu a}{c^2}\right\}I\right]d\tilde{V} = 0. \qquad (8.2.12)$$

Denoting by Δ_r, Δ_μ and Δ_ν respectively the space, angle and frequency differences, the discrete form of equation (8.2.12) for a hypersurface $d\tilde{V} = \Delta V \Delta\mu \Delta\nu$ can be written as

$$\frac{\delta}{\delta t}\left[\left(\frac{I}{c}\right) \cdot \Delta V \Delta\mu\Delta\nu\right] + \left\{\Delta_r\left[r^2 v_{\text{rel}}\left(\frac{I}{c}\right) + r^2 \mu I\right]\right\}\Delta\mu\Delta\nu$$
$$+ \left\{\Delta_\mu\left[-\dot{\mu}\left(\frac{I}{c}\right) + (1-\mu^2)\frac{1}{r} + \frac{\mu}{c}\left(\frac{3v}{r} - \frac{\partial(r^2 v)}{\partial \text{Vol}} - \frac{a}{c^2}I\right)\right]\right\}$$
$$\times \Delta V \Delta\nu - \left\{\Delta_\nu\left[\dot{\nu}\left(\frac{I}{c}\right) + \nu(1-3\mu^2)\frac{v}{cr} + \frac{\mu^2}{c}\frac{\partial(r^2 v)}{\partial \text{Vol}} + \frac{\mu a}{c^2}\right.\right.$$
$$\left.\left.I\right]\right\}\Delta V \Delta\mu + \left[-\eta + \left\{\chi + (1-3\mu^2)\frac{v}{cr} + \frac{\mu^2}{c}\frac{\partial(r^2 v)}{\partial \text{Vol}}\right.\right.$$
$$\left.\left. + \frac{2\mu a}{c^2}\right\}I\right]\Delta V \Delta\mu\Delta\nu = 0. \qquad (8.2.13)$$

The above equations are not used in the present form to solve the problem. One usually takes the zeroth and first moment against the angle variable μ before solving the equations. But this leads to two equations in three unknowns namely the radiation energy, flux and pressure. To link these variables the Eddington factor

$$f = K/J$$

is used. In the present context these factors have to be determined reasonably well. A wrong behaviour assumed for f can lead to a complete disaster. The Eddington factor is usually computed from a simplified form of equation (8.2.13). We will now obtain the two moment equations.

The radiation energy $E(\nu)$, flux $\bar{F}(\nu)$ and pressure $P(\nu)$ defined in equations (1.1.16), (1.1.21) and (1.1.28) respectively are related to the moments J, F and K by the relations

$$E(\nu) = \frac{4\pi}{c}J(\nu); \quad \bar{F}(\nu) = \pi F(\nu); \quad P(\nu) = \frac{4\pi}{c}K(\nu).$$

Taking the zeroth and first moment of the equation (8.2.12) against the angle μ, we obtain the first two moment equations. Integrating the resulting equations against the frequency ν we obtain the equations for the frequency integrated radiation energy E, flux $\bar{F}$ and pressure P. They are given by

8. Numerical Methods for Transfer Equations in CMF

$$\frac{d}{dt}\left[\int_V E d\text{Vol}\right] + \int_{\partial V} E(\mathbf{V}_{\text{rel}} \cdot d\mathbf{S}) + \int_V \frac{\partial}{\partial \text{Vol}}(r^2 \bar{F}) d\text{Vol}$$
$$+ \int_V \left[(E - 3P)\frac{v}{cr} + P\frac{\partial(r^2 v)}{\partial \text{Vol}} + \frac{2a\bar{F}}{c^2}\right] d\text{Vol}$$
$$+ \int_V \left[\int_0^\infty \{-4\pi\eta(\nu) + c\chi(\nu)E(\nu)\} d\nu\right] d\text{Vol} = 0 \tag{8.2.14}$$

and

$$\frac{d}{dt}\left[\int_V \frac{\bar{F}}{c^2} d\text{Vol}\right] + \int_{\partial V} \frac{\bar{F}}{c^2}(\mathbf{V}_{\text{rel}} \cdot d\mathbf{S})$$
$$+ \int_V \left[\frac{\partial P}{\partial r} + \frac{(3P - E)}{r} + \frac{\bar{F}}{c^2}\frac{\partial v}{\partial r} + \frac{a}{c^2}(E + P)\right] d\text{Vol}$$
$$+ \int_V \left[\frac{1}{c}\int_0^\infty \{\chi(\nu)\bar{F}(\nu)\} d\nu\right] d\text{Vol} = 0. \tag{8.2.15}$$

The above equations of radiative transfer are solved together with the following flow equations. These are the mass equation

$$dm - \rho \, d\text{Vol} = 0; \tag{8.2.16}$$

and the continuity equation

$$\frac{d}{dt}\left[\int_V \rho d\text{Vol}\right] + \int_{\partial V} \rho \mathbf{V}_{\text{rel}} \cdot d\mathbf{S} = 0. \tag{8.2.17}$$

The discrete form of the equations (8.2.16),(8.2.17),(8.2.14), (8.2.15) and (8.2.1) respectively can be written as follows :

$$\Delta m - \rho \Delta \text{Vol} = 0 \tag{8.2.18a}$$

$$\frac{\delta}{\delta t}[\rho \Delta \text{Vol}] + \Delta(r v_{\text{rel}} \rho) = 0 \tag{8.2.18b}$$

$$\frac{\delta}{\delta t}\left[\frac{E}{\rho}\Delta\xi\right] - \Delta\left[\frac{\delta m}{\delta t}(\frac{E}{\rho}) - r\bar{F}\right]$$
$$+ P\Delta(rv) + \left[(E - 3P)\frac{v}{r} + \frac{2a\bar{F}}{c^2}\right]\Delta\text{Vol}$$
$$= [4\pi k_p B - c k_E E]\Delta\xi \tag{8.2.18c}$$

$$\frac{\delta}{\delta t}\left[\frac{\bar{F}}{\rho c^2}\Delta\xi\right] - \Delta\left[\frac{\delta m}{\delta t}\left(\frac{\bar{F}}{\rho c^2}\right)\right] + r\left(\Delta P + \frac{\bar{F}}{c^2}\Delta v\right)$$
$$+ \left[\frac{(3P-E)}{r} + \frac{a}{c^2}(E+P)\right]\Delta\text{Vol}$$
$$= -\frac{k_F}{c}\bar{F}\Delta\xi \qquad (8.2.18d)$$

$$-\sigma_g\frac{\delta r}{\delta t} + \Delta[f^r + f^s] = 0. \qquad (8.2.18e)$$

In the above equations $\xi = \rho\Delta\text{Vol}$, B is the Planck function and k_P, k_E and k_F are the Planck mean, the absorption mean and the flux mean given by

$$k_P = \int_0^\infty k^a(\nu)\frac{B(\nu)}{B}d\nu$$

$$k_E = \int_0^\infty k^a(\nu)\frac{E(\nu)}{E}d\nu$$

$$k_F = \int_0^\infty (k^a(\nu) + k^s(\nu))\frac{\bar{F}(\nu)}{F}d\nu.$$

$k^a(\nu)$ and $k^s(\nu)$ are the true absorption and scattering coefficients given by the relations

$$\chi(\nu) = \rho(k^a(\nu) + k^s(\nu))$$
$$\eta(\nu) = \rho(k^a(\nu) + k^s(\nu)) \cdot S(\nu)$$

with
$$S(\nu) = \frac{k^s(\nu)J(\nu) + k^a(\nu)B(\nu)}{k^a(\nu) + k^s(\nu)}$$

Winkler et al. (1983) have replaced the product $(rv_{\text{rel}}\rho)$ by the term $(-\frac{\delta m}{\delta t})$ in the difference equations to avoid convergence difficulties.

Thus we have five equations in (8.2.18) for the six dependent variables $r, m, \rho, E, \bar{F}$ and P. The extra relation needed is the Eddington factor which is computed from a simplified form of the equation (8.2.13). The present adaptive mesh scheme is found to be computationally efficient.

8.3 The Method of Short Characteristic Rays

In the iterative scheme of solving the radiative transfer equation, one usually starts with an assumed source function and improves the convergence of this function to its true value by repeated iteration. Once the source function is assumed, then the transfer equation can be viewed as a first order partial differential equation for the intensity. With this interpretation one can now define the characteristics of the p.d.e. In the present method this view is pursued and the intensity is calculated along the characteristics between neighbouring spatial grid points defined on the spherical atmosphere. This is the reason for calling this the short characteristic method. In the present scheme this method is combined with the operator perturbation method discussed in chapters 5 and 8 to solve the transfer problem of the expanding spherical atmosphere. Kunasz and Auer (1987) and later Olson and Kunasz (1987) used the method of short characteristic rays to solve the transfer equations in the plane media. Recently Hauschildt (1992) has extended the method for expanding spherically symmetric atmospheric envelopes. In the comoving frame the time independent transfer equation can be written as [cf. (2.5.18) with the suffix zero suppressed and time dependent terms dropped].

$$\gamma(\mu+\beta)\frac{\partial I_\nu(r,\mu)}{\partial r}$$
$$+\frac{\partial}{\partial \mu}\left[\gamma(1-\mu^2)\left\{\frac{1+\beta\mu}{r} - \gamma^2(\mu+\beta)\frac{\partial \beta}{\partial r}\right\} I_\nu(r,\mu)\right]$$
$$-\frac{\partial}{\partial \nu}\left[\gamma\nu\left\{\frac{\beta(1-\mu^2)}{r} + \gamma^2\mu(\mu+\beta)\frac{\partial \beta}{\partial r}\right\} I_\nu(r,\mu)\right]$$
$$+\gamma\left[\frac{2\mu+\beta(3-\mu^2)}{r} + \gamma^2(1+\mu^2+2\beta\mu)\frac{\partial \beta}{\partial r}\right] I_\nu(r,\mu)$$
$$= \eta - \chi I_\nu(r,\mu). \tag{8.3.1}$$

In this equation, the matter velocity $v(r)$ is measured, as usual, in units of the speed of light $c \cdot \beta = v(r)/c$ and $\gamma = 1/\sqrt{1-\beta^2}$. For convenience this equation is written in terms of wave-length scale λ instead of the frequency ν in the form

8.3 The Method of Short Characteristic Rays

$$a_r \frac{\partial I}{\partial r} + a_\mu \frac{\partial I}{\partial \mu} + a_\lambda \frac{\partial (\lambda I)}{\partial \lambda} + 4a_\lambda I = \eta - \chi I, \qquad (8.3.2)$$

where

$$a_r = \gamma(\mu + \beta)$$

$$a_\mu = \gamma(1 - \mu^2)\left[\frac{1 + \beta\mu}{r} - \gamma^2(\mu + \beta)\frac{\partial \beta}{\partial r}\right]$$

$$a_\lambda = \gamma\left[\frac{\beta(1 - \mu^2)}{r} + \gamma^2\mu(\mu + \beta)\frac{\partial \beta}{\partial r}\right].$$

The characteristic rays of equation (8.3.2) are defined by

$$\frac{dr}{a_r} = \frac{d\mu}{a_\mu}.$$

Taking s as the geometrical path length along the characteristic ray, we rewrite the above equations as

$$\frac{dr}{ds} = a_r \qquad (8.3.3a)$$

$$\frac{d\mu}{ds} = a_\mu. \qquad (8.3.3b)$$

Along the characteristic ray, equation (8.3.2) becomes.

$$\frac{\partial I}{\partial s} + a_\lambda \frac{\partial (\lambda I)}{\partial \lambda} = \eta - (\chi + 4a_\lambda)I. \qquad (8.3.4)$$

Before we integrate along the characteristic ray we have to know the variations of μ along the ray in order to compute the term a_λ properly. We shall for the present restrict our attention to rays that are tangent to the spherical spatial grid of the atmosphere. These rays can be characterised by the impact parameter p, in our usual notation. At the point of tangency we set the path length s to be zero. Hence from (8.3.3a).

$$\mu|_{s=0} = -\beta(p).$$

In order to determine the variation of μ along the characteristics, we note that the variation of the cosine of the angle in the fixed lab frame μ_E is known. This is given by

$$\mu_E(r) = \pm\sqrt{(1 - p^2/r^2)}. \tag{8.3.5}$$

The corresponding cosine of the angle in the comoving frame μ can be obtained from μ_E using the Lorentz transformation. We have from (2.2.19)

$$\mu(r) = \frac{\mu_E(r) - \beta(r)}{1 - \beta(r)\mu_E(r)}. \tag{8.3.6}$$

Equations (8.3.5) and (8.3.6) show that the function $\mu(r)$ is purely a local function of r, i.e. μ depends on r through $\beta(r)$ and $\mu_E(r)$. Now we are in a position to compute the path length s along the ray. The characteristic equation (8.3.3a) is used for this. Using a Gauss-quadrature integration formula with linear interpolation for the velocity law between the radial points, one can obtain a fairly accurate path length s along the characteristic ray between the spatial grid points. Once the variation of μ along the path and the path length s is known, we can integrate equation (8.3.4) for the intensity. Hauschildt follows his earlier work (1991) and Mihalas' work (1980) to discretize the wave length derivative with a fully implicit scheme. This ensures stability. We have

$$\left.\frac{\partial(\lambda I)}{\partial \lambda}\right|_{\lambda=\lambda_k} = \frac{\lambda_k I_{\lambda_k} - \lambda_{k-1} I_{\lambda_{k-1}}}{\lambda_k - \lambda_{k-1}}.$$

Using this in equation (8.3.4) we obtain

$$\frac{dI_{\lambda_k}}{ds} + a_\lambda \frac{\lambda_k I_{\lambda_k} - \lambda_{k-1} I_{\lambda_{k-1}}}{\lambda_k - \lambda_{k-1}} = \eta_{\lambda_k} - (\chi_{\lambda_k} + 4a_\lambda) I_{\lambda_k}. \tag{8.3.7}$$

At this stage we define a new optical scale τ along the characteristic ray as follows:

$$-d\tau = \left[\chi_{\lambda_k} + a_\lambda\left(4 + \frac{\lambda_k}{\lambda_k - \lambda_{k-1}}\right)\right] ds$$
$$= \hat{\chi} ds, \text{ say}. \tag{8.3.8}$$

Introducing a modified source function

$$\hat{S} = \frac{\chi}{\hat{\chi}}\left[S + \frac{a_\lambda}{\chi}\left(\frac{\lambda_{k-1}}{\lambda_k - \lambda_{k-1}}\right) I_{\lambda_{k-1}}\right],$$

8.3 The Method of Short Characteristic Rays

where the original source function is $S = \eta/\chi$; we can rewrite (8.3.7) as

$$\frac{dI}{d\tau} = I - \hat{S}. \tag{8.3.9}$$

With this definition of optical depth and source function, the formal solution along the characteristics can be written in the following form :

$$I_i^- = I_{i-1}^- \exp(-\Delta\tau_{i-1}) + \int_{\tau_i}^{\tau_{i-1}} \hat{S}(t) \exp(-(t-\tau_i)) dt, \quad s < 0; \tag{8.3.10a}$$

$$I_i^+ = I_{i+1}^+ \exp(-\Delta\tau_{i-1}) + \int_{\tau_i}^{\tau_{i+1}} \hat{S}(t) \exp(-(t-\tau_i)) dt, \quad s > 0. \tag{8.3.10b}$$

The index i represents the discretization of the radial coordinate with $r_1 = 0$ at the outer boundary. τ_i are the corresponding optical depth of the spatial grids along the characteristics with $\tau_1 = 0$ and $\tau_{i-1} < \tau_i$. Also $\Delta\tau_{i-1} = \tau_{i-1} - \tau_i$. The integrations along each ray starts at the point furthest from the observer ($s < 0$) and proceeds towards the observer ($s > 0$). If we label the ray tangent to the shell $(j + 1)$ by the index j and introduce the optical thickness $\Delta\tau_{i-1,j}$ along the ray j between the spatial grid points τ_i and τ_{i-1} we have

$$\Delta\tau_{i-1,j} = \frac{1}{2}[\tilde{\chi}_{i-1} + \tilde{\chi}_i]\left|s_{i,j} - s_{i-1,j}\right|. \tag{8.3.11}$$

Equation (8.3.10) can be now written as

$$I_{i,j}^- = I_{i-1,j}^- \exp(-\Delta\tau_{i-1,j}) + \Delta I_{i,j}^- \quad s < 0; \tag{8.3.12a}$$

$$I_{i,j}^+ = I_{i+1,j}^+ \exp(-\Delta\tau_{i-1,j}) + \Delta I_{i,j}^+ \quad s > 0. \tag{8.3.12b}$$

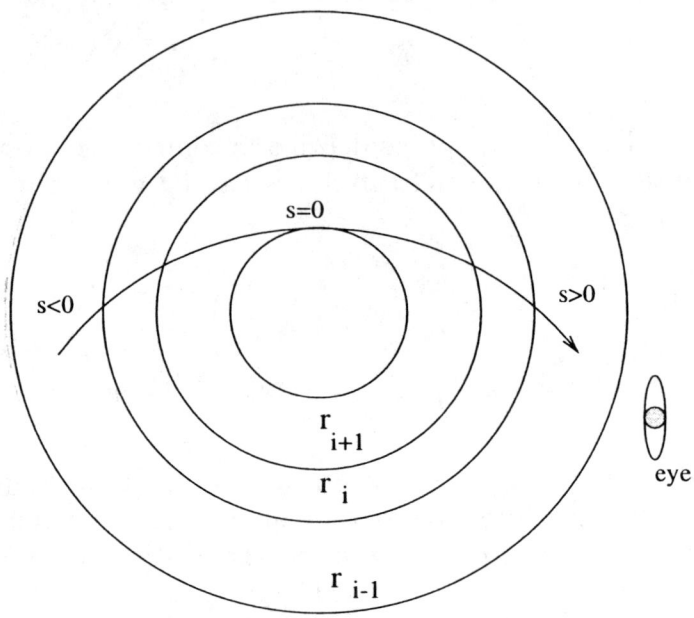

Figure 8.3. Direction of tangential characteristic ray in a spherical envelope

We shall suppress the ray index j for convenience and rewrite the above equations as

$$I_i^- = I_{i-1}^- \exp(-\Delta\tau_{i-1}) + \Delta I_i^-, \quad s < 0; \qquad (8.3.13a)$$

$$I_i^+ = I_{i+1}^+ \exp(-\Delta\tau_{i-1}) + \Delta I_i^+, \quad s > 0; \qquad (8.3.13b)$$

In the computation of the increments $\Delta I_i^\pm$ of the intensity, the source function $\hat{S}$ along a characteristic ray is interpolated by either a linear or parabolic polynomial so that

$$\Delta I_i^\pm = \alpha_i^\pm \hat{S}_{i-1} + \beta_i^\pm \hat{S}_i + \gamma_i^\pm \hat{S}_{i+1}.$$

The coefficients $\alpha_i, \beta_i, \gamma_i$ can be calculated [cf. Hauschildt (1992), equations (24), (25)]. For instance in the case of linear interpolation we have

$$\alpha_i^- = e_{0,i} - e_{1,i}/\Delta\tau_{i-1}$$
$$\beta_i^- = e_{1,i}/\Delta\tau_{i-1}$$
$$\gamma_i^- = 0$$
$$\alpha_i^+ = 0$$
$$\beta_i^+ = e_{1,i+1}/\Delta\tau_i$$
$$\gamma_i^+ = e_{0,i+1} - e_{1,i+1}/\Delta\tau_i$$

where

$$e_{0,i} = 1 - \exp(-\Delta\tau_{i-1})$$
$$e_{1,i} = \Delta\tau_{i-1} - e_{0,i}.$$

It must be noted that for spherical envelopes one must distinguish between the disc rays (i.e. rays that intersect the inner boundary) and the shell rays (i.e. rays that do not intersect the inner boundary). For the disc ray the integration is from the point 1 to point N followed by integration from point N+2 to point 2N where I_{N+1} is given as the inner boundary condition.

8.4 Operator Perturbation Method: Approximate Lambda Iterative Method for Non-LTE Radiative Transfer in Spherically Symmetric Expanding Atmospheres

We have outlined in chapter 3 Sobolev's escape probability method (1957) and in chapters 6 and 7 solutions of radiative transfer equations in comoving frame (CMF) for spherically symmetric moving media. Mihalas and Kunasz (1978) solved the coupled transfer equations in CMF and the rate equations for explaining the abnormal spectral line structure of Wolf–Rayet stars, possessing spherically symmetric expanding atmospheres. Hamann (1985) and other workers used this technique to explain the features of some special lines of such stars. However it was found that the execution of their scheme was rather time consuming.

Recently an operator perturbation method has been proposed for obtaining fast and accurate solutions of radiative transfer equations with constraints in non-LTE conditions both in static and moving media. This is done by the use of approximate integral and differential operators for solving coupled transfer and rate equations, estimating the errors arising out of these approximations and finally minimising the errors by suitably chosen iterative processes.

Our main interest in this treatise is to introduce basic mathematical and numerical approaches to study the transfer problems in spherically symmetric moving media. We shall trace below the brief history of the development of the approximate lambda opertor (ALO) and accelerated lambda iterative method (ALI). Schemes presented in the first part of section (8.4) were mostly involved with plane medium under static conditions. Although the methods developed there do not directly concern with the topic of the book, the knowledge of these basic structures will help to understand the use of the perturbation operator methods for solving the problem of our interest.

ALI for Plane Static Atmospheres

In early years, the complete linearisation technique of Auer and Mihalas [cf. Mihalas (1978), p. 230, 231] has been widely used to solve the coupled radiative transfer equations in non-LTE situations. In many cases the equations are nonlinear. The scope of the use of the method was broadened by Rybicki (1971) and Kalkofen (1974). The Newton–Raphson linearisation scheme [cf. Mihalas (1978), p. 117–119] used by them is highly efficient in terms of economy in the number of iterations involved. However, it was rather poor in the use of computer time and storage.

Cannon (1973a, 1973b) proposed a new method in which radiative transfer and statistical equilibrium equations are coupled into a single integral equation. The exact integral operator (Λ) of this equation is expressed in terms of a simple approximate (Λ^*) operator (ALO). Cannon divides the transfer problem into two separate parts of computation. First a solution, arising out of the use of the approximate operator in place of the exact, is calculated with just a few quadrature points. Then the error made by this solution in satisfying the constraint or conservation equations is determined very accurately with a dense quadrature set. The computation of this is not too time consuming. The basic feature of Cannon's approach is the perturbation of the exact integral operator about a simple approximate lambda operator (ALO), and expanding the source function in a series [cf. Kalkofen (1984), p. 428–431] and adopting an iterative scheme to solve the problem.

Scharmer (1981, 1984), Scharmer and Nordlund (1982), Scharmer and Carlsson (1985) put Cannon's ideas into a concrete shape. The basis of their method is a highly simplified description of the radiative transfer problem, which in turn led to the use of diagonal operators. The line transfer problem is formulated in terms of integral equation for complete redistribution in conservative form. The source function is linearised about an initial value. The errors made in the constraint or conservation equations of statistical equilibrium due to the adoption of certain order of the source function are corrected. An approximate Λ-operator is used to write down the exact solution as a perturbation series. The outgoing intensities are taken to be related with the source function by modified Eddington–Barbier relations both in the core and the wings of the line [cf. chapter 5]. The inward intensities due to internal sources are assumed to be zero. With these approximations the Λ matrix takes nearly a triangular shape and can be built up very rapidly. Scharmer's approach included in it basic ideas of Cannon's operator perturbation, Rybicki's core saturation [cf. chapter 5, section 3] and Sobolev's escape probability method [cf. chapter 3]. An outline of Scharmer's operator perturbation technique is given by Kalkofen ((1984) p. 431–433).

A host of other methods, more or less based on the same principles like Cannon's and Scharmer's perturbation operator method, were also introduced about the same time. We may mention the works of Hamann (1985, 1986), Werner and Husfield (1985), Werner (1986, 1987), Olson, Auer and Buchler (1986). We summarise below the fundamental features of the present accelerated lambda iteration as against the classical lambda iteration. For simplicity, we consider only the continuum radiative transfer.

The formal solution of continuum radiative transfer in *classical Λ-iteration method* is given by

$$J_\nu = \Lambda_\nu S_\nu, \qquad (8.4.1)$$

where J_ν is the mean intensity, S_ν, the continuum source function and Λ_ν the lambda operator.

However, in actual practice, the matrix equation to be solved for the transfer problem can be written as

$$S_\nu = \mathbb{T}_\nu J_\nu, \qquad (8.4.2)$$

where $\mathbb{T}_\nu$ is the transfer matrix and

$$\Lambda_\nu = (\mathbb{T}_\nu)^{-1}. \qquad (8.4.3)$$

After all the variables are discretised, $\mathbb{T}_\nu$ takes a tridiagonal form [cf. Mihalas (1978)]. The transfer equation is then solved by Feautrier's method [cf. Feautrier (1964)].

The idea of separating a linear operator like Λ into a sum of two parts, one of which has a simple inverse, and using this as the basis for an iteration scheme is well known in numerical analysis. Cannon introduced this in radiative transfer studies. An approximate lambda operator Λ^* is chosen and the initial lambda operator Λ is split by writing

$$\Lambda = \Lambda^* + (\Lambda - \Lambda^*)$$

and equation (8.4.1) is rewritten as

$$\begin{aligned} J_\nu^{\text{new}} &= \Lambda_\nu^* S_\nu^{\text{new}} + (\Lambda_\nu - \Lambda_\nu^*) S_\nu^{\text{old}} \\ &= \Lambda_\nu^* S_\nu^{\text{new}} + \triangle J_\nu. \end{aligned} \qquad (8.4.4)$$

$\triangle J_\nu$ is the correction term, S_ν^{old} denotes the source function based on population values secured at the end of the previous operation and S_ν^{new}, those to be used at the start of the new iteration. In the works of Hamann, Werner and Husfield, Λ^* is taken to be a diagonal operator, which acts locally. This implies that the statistical equation depends only on local neighbourhood conditions. The non-local condition of the radiation field is given by $\triangle J_\nu$. Olson et al. (1986) show that

$$\Lambda_\nu^* = diag[\mathbb{T}_\nu]^{-1}, \qquad (8.4.5)$$

is an ideal choice for a strictly local ALO. This choice of ALO [cf. Olson et al. (1986)] results in significantly faster convergence of calculation process saving considerable amount of computer time.

Now in what follows, we describe the use of ALI for studying the problem of line formation in a spherically symmetric expanding atmosphere which is the topic of this book.

ALI for Line formation in Static and Expanding Atmospheres

We outline below the use of ALI for studying the problem of line formation in expanding atmospheres. To do this, we start with the work of Hamann (1978), considering for simplicity the statistical equilibrium equation for two-level atoms. However, Hamann (1987) extended his algorithm to the study of multi-level atoms as well.

We recall [cf. equation (5.2.2)] that the formal solution of the transfer equation relating the mean intensity $\bar{J}$ and the source function S is given by

$$\bar{J} = \int K(\mathbf{r},\mathbf{r}')\chi(\mathbf{r}')S(\mathbf{r}')d^3\mathbf{r}'.$$

The kernel $K(\mathbf{r},\mathbf{r}')$ depends on the geometry and the nature of the atmosphere. We usually denote this by

$$\bar{J} = \Lambda S, \tag{8.4.6}$$

where Λ is called the lambda operator. Also the statistical equilibrium equation for the standard two level atom model with complete redistribution [cf. (5.1.10)] can be put in the form

$$S = (1-\epsilon)\bar{J} + \epsilon B. \tag{8.4.7}$$

Eliminating the source function from the two equations, we have

$$\bar{J} = \Lambda[(1-\epsilon)\bar{J} + \epsilon B],$$

i.e.

$$\bar{J} = \Lambda(1-\epsilon)\bar{J} + \Lambda\epsilon B. \tag{8.4.8}$$

A new approximate lambda operator Λ^* is introduced and the equation (8.4.6) is written as

$$\bar{J} = (\Lambda - \Lambda^*)S + \Lambda^* S. \tag{8.4.9}$$

The following iterative scheme is then proposed to solve the above equation.
$$\bar{J}_{\text{new}} - \Lambda^* S_{\text{new}} = (\Lambda - \Lambda^*) S_{\text{old}}. \tag{8.4.10}$$
The above scheme becomes viable provided

(a) Λ^* can be inverted easily so that $\bar{J}_{\text{new}}$ and S_{new} can be consistently determined from (8.4.6). The operator Λ^* is usually taken to be diagonal for easy inversion.

(b) $(\Lambda - \Lambda^*)$ does not yield contributions from optically thick line cores.

In the standard lambda iterative scheme, the largest amplification matrix in the case of Doppler-profiles is approximately of the order $(1-\epsilon)(1-\tau_0^{-1})$ where τ_0 is the optical thickness of the medium. For small ϵ and large τ_0 this is very close to unity and, therefore, the convergence rate is very poor. Although the choice $\Lambda^* = \Lambda$ would be the best, this is not helpful for (a) and hence Λ^* is constructed to be as close as possible to either the diagonal or tridiagonal of Λ operator. This improves the convergence rate. Substituting (8.4.7) in (8.4.10), we have

$$\bar{J}_{\text{new}} - \Lambda^*[(1-\epsilon)\bar{J}_{\text{new}} + \epsilon B]$$
$$= (\Lambda - \Lambda^*)[(1-\epsilon)\bar{J}_{\text{old}} + \epsilon B]$$
$$= \Lambda[(1-\epsilon)\bar{J}_{\text{old}} + \epsilon B] - \Lambda^*(1-\epsilon)\bar{J}_{\text{old}} - \Lambda^*\epsilon B$$
$$= \Lambda S_{\text{old}} - \Lambda^*(1-\epsilon)\bar{J}_{\text{old}} - \Lambda^*\epsilon B.$$

Denoting ΛS_{old} by $\bar{J}_{Fs}$, we have

$$\bar{J}_{\text{new}} - \Lambda^*(1-\epsilon)\bar{J}_{\text{new}} = \bar{J}_{Fs} - \Lambda^*(1-\epsilon)\bar{J}_{\text{old}}.$$
$$\bar{J}_{\text{new}} = [1 - \Lambda^*(1-\epsilon)]^{-1}[\bar{J}_{Fs} - \Lambda^*(1-\epsilon)\bar{J}_{\text{old}}]. \tag{8.4.11}$$

It must be noted that in the above iterative scheme we have introduced $\bar{J}_{Fs}$ which is infact the formal solution of equation (8.4.6) with the old source function. The main attraction of the above iterative scheme is that this formal solution $\bar{J}_{Fs}$ need not be very accurate and a fast crude approximation is sufficient for a good convergence of the scheme. Further there is enough flexibility in the choice of the approximate operator Λ^* subject to the two conditions being satisfied. A very simple form for Λ^* suggested by Rybicki is based on the core saturation idea and is given by

$$\Lambda_\nu^* S = \begin{cases} S, & \text{in optically thick line core} \\ 0, & \text{elsewhere.} \end{cases} \qquad (8.4.12)$$

The nature of this diagonal operator is based on the monochromatic Λ-operator given by

$$\Lambda_\nu(\tau_\nu, \tau_\nu') = \begin{cases} \delta(\tau_\nu, \tau_\nu'), & \tau_\nu > \gamma \\ 0, & \text{otherwise.} \end{cases}$$

γ is the optical depth to the nearest boundary. It is an adjustable parameter which suggests that in the core-wing approximation, contribution to $\bar{J}$ comes only from frequency-angle values for which $\tau_{\nu\mu} > \gamma$.

We introduce the core fraction f_c and rewrite the above operator as

$$\Lambda_\nu^* = f_c S, \qquad (8.4.13)$$

where

$$f_c = \begin{cases} 1, & \text{in optically thick line core} \\ 0, & \text{elsewhere.} \end{cases} \qquad (8.4.14)$$

Thus equation (8.4.11) becomes

$$\bar{J}_{\text{new}} = [1 - (1-\epsilon)f_c]^{-1}[\bar{J}_{Fs} - (1-\epsilon)f_c \bar{J}_{\text{old}}].$$

i.e.

$$\bar{J}_{\text{new}} - \bar{J}_{\text{old}} = [1 - (1-\epsilon)f_c]^{-1}[\bar{J}_{Fs} - \bar{J}_{\text{old}}]. \qquad (8.4.15)$$

Equation (8.4.15) specifies an accelerated Λ-iteration process, $[1 - (1-\epsilon)f_c]^{-1}$ acting as the amplification factor.

Core Fraction in Spherically Symmetric Expanding Atmospheres

The above iterative scheme will be well defined once the core fraction f_c is known. We will now discuss how this fraction is determined in the moving spherical atmosphere. As the profile function is isotropic in the comoving frame CMF we usually compute f_c in this frame. As the core fraction f_c denotes that part of the scattering integral which falls within the optically thick line core, the task now is to determine at each depth point the frequency range $(x_{\text{red}}, x_{\text{blue}})$ of the optically thick line core. We define

$$f_c = \frac{\int_{x_{\text{red}}}^{x_{\text{blue}}} \phi dx}{\int_{-x_{\text{max}}}^{x_{\text{max}}} \phi dx}, \qquad (8.4.16)$$

where we have introduced the exact normalization for f_c with respect to a finite band width $(-x_{\text{max}}, x_{\text{max}})$ usually set to be 3 or 4 Doppler widths. A parameter γ of order unity is used to test whether a given frequency belongs the core or not. If there exists a ray starting from the given point which reaches the boundary in an optical depth shorter than γ, then this frequency is outside the core- confining frequencies $(x_{\text{red}}, x_{\text{blue}})$. To avoid a large number of computations, two simplifying assumptions were made by Hamann (Kalkofen, (1987), p. 43.).

(a) Only two spatial directions, namely, the radial and the transverse rays were considered.
(b) The velocity gradient and opacity were assumed to be constant in the vicinity of the point under consideration.

Consider a photon traveling along a ray with a given impact parameter p across the spherically expanding atmosphere. It will have exact resonance with the line absorbing atoms exactly at one spatial point r. The actual interaction may take place within a finite range around this exact resonance point due to the width of the absorption profile in the moving frame. The point r will be considered as ' internal point' if this spatial range lies entirely inside the boundaries of the atmosphere. For such points the computation of optical depth by integration over spatial distance can be transformed into an integration over frequencies relative to the CMF. As any two points in the expanding atmosphere are receding from one another, the frequency in the rest frame is always red-shifted in the CMF irrespective of the spatial direction of the ray. If $v'(r, p)$ is the projected velocity gradient on the given ray, then by the second assumption of constancy in this approximation, the optical depth is given by

$$\tau(r, p) = \frac{\chi(r)}{v'(r, p)}. \qquad (8.4.17)$$

For the two spatial directions considered, the projected velocity gradient is given by

$$v'(r,p) = \begin{cases} \frac{dv}{dr}, & \text{radial direction} \\ \frac{v}{r}, & \text{transverse direction.} \end{cases} \quad (8.4.18)$$

The above method of determining the core range is valid only for interior points, where it has been tacitly assumed that

$$x_{\text{red}} - (-x_{\text{max}}) < \Delta v, \quad (8.4.19)$$

Δv is the relative velocity between the point considered and the boundary point. This depends also on the direction of the ray. If the condition (8.4 19) is not satisfied, then we are in the so-called "boundary zone". In this case the x_{red} for the core-range must satisfy

$$\Phi(x_{\text{red}}) - \Phi(x_{\text{red}} - \Delta v) = \frac{\gamma}{\tau}, \quad (8.4.20)$$

where

$$\Phi(x) = \int_{-x_{\text{max}}}^{x} \phi dx.$$

Clearly (8.4.20) has a solution provided

$$\Phi\left(\frac{\Delta v}{2}\right) - \Phi\left(-\frac{\Delta v}{2}\right) > \frac{\gamma}{\tau}, \quad (8.4.21)$$

assuming the profile is symmetrical about the centre and monotonically decreasing towards the wings. If (8.4.21) is not satisfied, then we set $f_c = 0$. We take the smaller of the two solutions for x_{red} from equation (8.4.20) and define

$$x_{\text{blue}} = \Delta v - x_{\text{red}} \quad (8.4.22)$$

to obtain the core-range.

In this approximation, the above procedure is performed for both the radial and transverse directions at each point and the minimum extension of the core is taken to estimate the core-range.

Hamann's approach aims at developing an accelerated Λ-iteration process using an approximate Λ-operator which depends on core fraction f_c and a free parameter γ which is adjusted to secure maximum convergence rate. γ divides the core of the line from the wings.

Hamann (1987) used the ALI scheme to study the line formation problems in expanding atmospheres consisting of multi-level atoms as well. In non-LTE multi-level problems, the rate coefficients of the statistical equilibrium equations are influenced by the radiation field and the population numbers in rate equations depend non-linearly on the radiation field. Hamann (1987, p. 52–55, 63–65) developed a lambda iteration scheme with suitably defined approximate operator to obtain modified rate equations at each depth point. ALI is done by a procedure formally similar to that in the case of two-level atoms. The radiative transfer equation is solved in the comoving frame (CMF). The constraint of radiative equilibrium is not taken into account.

In developing the contemporary methods based on the same principle, attention is always paid to saving computer time in their execution. On this account, Hamann's approach is considered rather slow (Hillier, (1990), p. 117) even when accelerated by Ng's (1974) scheme. Olson, Auer and Buchler (1986), Olson and Kunasz (1987), and Puls and Herrero (1988) studied the accelerated lambda iteration technique from a mathematical point of view and constructed approximate lambda operators ALO to obtain an optimal rate of convergence. A comparative analysis of a number of ALO's is made by Puls and Herrero (1988). Hempe and Schönberg (1986) and Schönberg and Hempe (1986) derived a useful diagonal operator, linearising the transfer equation under the assumption of locally controlled intensity. Hamann and Schmutz (1987) avoided using the constraint equation of radiative equilibrium in studying the problem of line formation in Wolf–Rayet stars. They assumed instead temperature structures evaluated in grey LTE approximation. Similar approach was followed by Hamann et al. (1988) and Schmutz et al. (1987, 1989). Pauldrach and Herrero (1988), in their discussion of ALI techniques for modeling multi-level atoms in optically thick "unified" atmospheres (photosphere plus wind), used a temperature structure rather than the constraint equation of radiative equilibrium.

Hamann and Wessolowski (1989) set out to improve the treatment of radiative equilibrium accounting for the conservation of energy in non-grey non-LTE radiative transfer. We outline below their treatment of line formation in expanding atmospheres

8.4 Operator Perturbation Method

based on the study of radiative equilibrium by ALO. We follow Hamann's (1987) formulation of iteration with ALO described earlier. The radiative field $\mathbf{J}_{Fs}$ is obtained from the formal solution of the transfer equation written in integral form, starting from a set of old population numbers n_{old} and the non-LTE source function S_{old}. From equation (8.4.10) we may write

$$\bar{J}_{\text{new}} = \bar{J}_{Fs} + \Lambda^*(S_{\text{new}} - S_{\text{old}}). \tag{8.4.23}$$

where

$$\bar{J}_{Fs} = \Lambda S_{\text{old}}. \tag{8.4.24}$$

Λ^* is an ALO, which is chosen to act 'locally' to avoid coupling of rate equations in space.

Then we write the equations of statistical equilibrium as

$$\mathbf{n}_{\text{new}} \cdot \mathbb{P}(\bar{J}_{\text{new}}) = \mathbf{C}, \tag{8.4.25}$$

where $\mathbb{P}$ is the rate coefficient matrix.

For each depth point, the population number is written as a row vector

$$\mathbf{n} = (n_1, n_2, \ldots, n_N). \tag{8.4.26}$$

Thus (8.4.25) represents a set of algebraic equations at each spatial point. $\mathbb{P}$ itself depends on $\mathbf{n}_{\text{new}}$, so that (8.4.25) is non-linear. We write these N-equations in the form

$$f(\mathbf{n}_{\text{new}}) = [f_1, f_2, \ldots, f_N] = \mathbf{n}_{\text{new}} \cdot \mathbb{P} - \mathbf{C} = 0. \tag{8.4.27}$$

The corrections for population numbers in the iteration step from k to $k+1$ is given by Newton-Raphson iteration process

$$\triangle \mathbf{n} = \mathbf{n}^{k+1} - \mathbf{n}^k = -f(\mathbf{n}^k) \cdot \mathbb{M}^{-1}, \tag{8.4.28}$$

where the elements of the matrix $\mathbb{M}$ is given by

$$\mathbb{M}_{ij} = \frac{\partial f_j(\mathbf{n})}{\partial n_i}\bigg|_{\mathbf{n}=\mathbf{n}^k}. \tag{8.4.29}$$

If we now introduce the constraint of radiative equilibrium, the temperature T becomes an additional independent variable and the row vector $\mathbf{n}$ changes to $[\mathbf{n}, T]$. Because of the dependence of rate equations on T, the linearised matrix $\mathbb{M}$ has an additional row containing the temperature derivatives, the elements of which are

$$\bar{M}_{i=T,j} = \frac{\partial f_j[\mathbf{n},T]}{\partial T}. \qquad (8.4.30)$$

Now we write the radiative equilibrium equation as

$$f_T(\mathbf{n},T) = 0. \qquad (8.4.31)$$

Then the corrections for the population, for an iterative solution of the linearised equations are

$$[\triangle \mathbf{n}, \triangle T] = [-f(\mathbf{n},T]), -f_T] \cdot \bar{M}^{-1}, \qquad (8.4.32)$$

where $\bar{M}$ is the matrix M further augmented by an additional column with

$$\bar{M}_{i,j=T} = \frac{\partial f_T[\mathbf{n},T]}{\partial n_j}, \quad (i = 1, 2, \ldots, N) \qquad (8.4.33)$$

and

$$\bar{M}_{i=T,j=T} = \frac{\partial f_T[\mathbf{n},T]}{\partial T}. \qquad (8.4.34)$$

Further the new equilibrium equation becomes

$$f_T(\mathbf{n}_{\text{new}}, T_{\text{new}}) = \int_0^\infty [\eta_{\text{new}} - \chi_{\text{new}} J_{\text{new}}] d\nu. = 0. \qquad (8.4.35)$$

In the case of rapidly expanding atmospheres, the solution for line radiative transfer is done in CMF and for continuum radiation, the static frame is used. For the radiative equilibrium, under the same approximation

$$f_T = \int_0^\infty [\eta_c - \chi_c J_c] d\nu + \sum_{\text{lines}} [\eta_L - \chi_L \bar{J}_L]. \qquad (8.4.36)$$

The subscripts c and L denote continuum and line respectively.

The rate of convergence of the process depends on the judicious choice of ALO. The convergence can be accelerated by Ng's (1974) method. Test calculations with WR-type of stars reveal that for spherically expanding atmospheres consisting mainly of helium, the temperature stratification obtained with radiative equilibrium equation included differs only moderately from the grey, LTE model used before.

Complete Linearisation Approach for Solving Radiative and Statistical Equilibrium Equations

Hillier's (1990) approach for solving radiative and statistical equilibrium equations in expanding atmospheres is based on the complete linearisation method of Auer and Mihalas (1969). He uses an improved form of *tridiagonal* Newton–Raphson operator devised by Schönberg and Hempe (1989). We outline below the basic elements of Hillier's method.

He assumes a given set of population numbers to solve the radiative transfer equation for all line and continuum frequencies. This set of population is expected to be inconsistent with the radiative field suggested by statistical equilibrium equations. New estimates of populations have to be made and this is done by the linearisation of equilibrium equations. However in the complete linearisation of non-LTE problems of radiative transfer with constraints of statistical equilibrium all variables are globally coupled. As a consequence, two factors have to be kept in mind.

(i) All variables interact and hence no variable can be considered more fundamental than the other.

(ii) The variables are strongly coupled through radiative transfer process. A change of any variable at any point of the medium is attended with change of all other variables at all other points.

The system of linearised equations is indeterminate and has to be solved by iteration.

The linearisation of constraint equations and radiative equations are done as follows [cf. Mihalas (1978), p. 230–233, Werner (1984), p. 74–85].

The system of linearised constraint equations (statistical equilibrium, charge conservation, etc) at each depth point d may be written as [cf. Werner (1987), p. 83]

$$\mathbb{M}_d \cdot \delta \Psi_d = \mathbf{C}_d, (d = 1, 2, \ldots . D), \qquad (8.4.37)$$

where

$\mathbb{M}_d,$ is a matrix of dimension $N_c \times N_T$

$N_c,$ is the number of constraint equations

N_T, number of variables describing the population and radiation field

$\mathbf{C}_d$, is a vector of length N_c containing the error in the constraint equations evaluated with the assumed populations and radiation field.

$$\delta \boldsymbol{\Psi}_d = (\delta n_1, \ldots, \delta n_D, \delta N_e, \delta T_d, \delta J_1, \ldots, \delta \bar{J}_1 \ldots), \qquad (8.4.38)$$

where n_d denotes occupation numbers, T the temperature and N_e the number of electrons. Let $f_d(\boldsymbol{\Psi}_d) = 0$ be the complete system of constraint equations with $\boldsymbol{\Psi}_d$ as it's solution. Let $\boldsymbol{\Psi}_d^0$ be the approximate solution due to the approximate population assumed. Then

$$\boldsymbol{\Psi}_d = \boldsymbol{\Psi}_d^0 + \delta \boldsymbol{\Psi}_d$$

where $\delta \boldsymbol{\Psi}_d$ is the perturbation to the current imperfect solution $\boldsymbol{\Psi}_d^0$. The perturbation $\delta \boldsymbol{\Psi}_d$ is suitably chosen so that $f(\boldsymbol{\Psi}_d^0 + \delta \boldsymbol{\Psi}_d) \to 0$, in the iteration. This perturbed $\delta \boldsymbol{\Psi}_d$ is obtained by linearising the original system, viz.

$$f_d(\boldsymbol{\Psi}_d^0) + \sum_j \frac{\partial f_d}{\partial \boldsymbol{\Psi}_{d,j}} \delta \boldsymbol{\Psi}_{d,j} = 0. \qquad (8.4.39)$$

The next step is to use linearised radiative transfer equation to express δJ in (8.4.38) in terms of $\delta n, \delta N_e$, and δT.

For this, the variables in the radiative transfer equation in spherically symmetric moving media are discretised by Feautrier's scheme (1964) and the difference equations of transfer (and boundary and initial conditions) are written out. For the solution of the system of equations Gaussian elimination scheme of Feautrier (or Rybicki) are used at every depth point [cf. chapters 6 and 7]. The system of equations can be written in the form [cf. (6.1.45); also Mihalas (1978), p. 507]

$$\mathbb{T}_n \mathbf{J}_n = \mathbf{X}_n + \mathbb{U}_n \mathbf{J}_{n-1} + \mathbb{V}_n \mathbf{H}_{n-1}, \qquad (8.4.40)$$

$$\mathbf{H}_n = \mathbb{A}_n \mathbf{J}_n + \mathbb{B}_n \mathbf{H}_{n-1}, \qquad (8.4.41)$$

where

8.4 Operator Perturbation Method

J, is the mean intensity vector

H, is the flux vector.

$\mathbb{T}$ is tridiagonal, $\mathbb{U}$, a diagonal and $\mathbb{V}$, a lower bidiagonal matrix. $\mathbb{A}$ is upper triangular matrix, $\mathbb{B}$, a diagonal matrix and $\mathbf{X}$ is a vector. The subscript n is the frequency index.

Hillier considered only line intensities. The frequency integrated mean intensity $\bar{J}$ is given by [cf. Hillier (1990), p. 118]

$$\bar{J} = \int_{-\infty}^{\infty} \phi_\nu J_\nu d\nu = \sum_{n=1}^{n_{max}} w_n J_n \phi_n, \qquad (8.4.42)$$

where w_n are quadrature weights for the frequency integral.

Now J and H are considered to be functions of four independent variables χ_c, S_c, χ_L, S_L where the symbols imply continuum opacity, continuum source function, line opacity and line source function respectively. Let x_j ($j = 1, 2, 3, 4$) represent these four independent variables.

At each depth point, we may write

$$\delta J_l = \sum_{d=1}^{D} \sum_{s=1}^{4} \frac{\partial J_l}{\partial x_{sd}} \delta x_{sd}. \qquad (8.4.43)$$

Equations (8.4.40) and (8.4.41) are linearised with respect to the independent variables and we get

$$\begin{aligned}\mathbb{T}_n \partial \mathbf{J}_n &= \mathbb{U}_n \partial \mathbf{J}_{n-1} + \mathbb{V}_n \partial \mathbf{H}_{n-1} \\ &+ [\partial \mathbf{X}_n - \partial \mathbb{T}_n \mathbf{J}_n + \partial \mathbb{U}_n \mathbf{J}_{n-1} + \partial \mathbb{V}_n \mathbf{H}_{n-1}]\end{aligned}$$
(8.4.44)

and

$$\partial \mathbf{H}_n = \mathbb{A}_n \partial \mathbf{J}_n + \mathbb{B}_n \partial \mathbf{H}_{n-1} + \partial \mathbb{A}_n \mathbf{J}_n + \partial \mathbb{B}_n \mathbf{H}_{n-1}. \qquad (8.4.45)$$

$\partial \mathbf{J}$ can be considered to be a three dimensional matrix with $\frac{\partial J_l}{\partial x_{sd}}$ as the elements.

From equations (8.4.42) and (8.4.43)

$$\delta \bar{J}_l = \sum_{d=1}^{D} \sum_{s=1}^{4} \partial \bar{J}_{lds} \, \delta x_{sd}. \qquad (8.4.46)$$

So far, the scheme followed is that of a modified complete linearisation process. At this stage an assumption is made that the radiation field is influenced mainly by local conditions and neighbourhood points. Hence non-local terms are ignored to obtain a local operator. The three-dimensional matrix $\partial \mathbf{J}$ can be replaced by $(\partial \mathbf{J}')$ ignoring the non-local terms in the variation matrices. The dimension of $(\partial \mathbf{J}')$ is $D \times N_B \times 4$ where N_B indicates the degree of depth coupling ($N_B = 1 \Rightarrow$ pure local variation; $N_B = 3 \Rightarrow \bar{J}$ depends on the independent variables locally and in its immediate neighbourhood).

We now linearise x in terms of unknown population numbers and write δx_{sd} at each depth point and for each independent variable as

$$\delta x_{sd} = \sum_{j=1}^{N_c} \left(\frac{\partial x_s}{\partial n_j} \right)_d \delta n_{jd}. \qquad (8.4.47)$$

N_c is the number of constraint equations.

Then from (8.4.43),(8.4.46) and (8.4.47)

$$\begin{aligned}
\delta \bar{J}_l &= \sum_{d=1}^{N_B} \sum_{s=1}^{4} \frac{\partial \bar{J}_l}{\partial x_{sr}} \delta x_{sr} \\
&= \sum_{d=1}^{N_B} \sum_{j=1}^{N_c} \sum_{s=1}^{4} \frac{\partial \bar{J}_l}{\partial x_{sr}} \left(\frac{\partial x_s}{\partial n_j} \right)_r \delta n_{jr} \\
&= \sum_{d=1}^{N_B} \sum_{j=1}^{N_c} \sum_{s=1}^{4} \delta \mathbf{J}'_{lds} \left(\frac{\partial x_s}{\partial n_j} \right)_r \delta n_{jr} \\
&= \sum_{d=1}^{N_B} \sum_{j=1}^{N_c} \delta \mathbf{J}''_{ldj} \delta n_{jr}. \qquad (8.4.48)
\end{aligned}$$

In (8.4.48) if $N_B = D, r = d$, otherwise $r = l - (N_0 + 1)/2 + d$. We can then replace $\delta \mathbf{J}$ by $\delta \mathbf{n}$ in equations (8.4.37) and (8.4.38) and hence they are now coupled at different depths through $\bar{J}$. An iterative scheme is used to lead to the populations suggested by statistical equilibrium equations to be consistent with the radiation field.

The choice of a good starting solution for the population is very crucial for this scheme to converge. Hillier (1990, p. 119) points out that the search for an adequate starting solution is by no means a trivial task. He suggested that locally computed Sobolev approximation may be used as a starting model for the CMF solution. The Sobolev approximation [cf. chapter 3] model may be used until it is sufficiently converged and then CMF formal solution can be started. However, there is certain amount of arbitrariness in the choice of switching point. He also noted that as a consequence of linearisation, the populations may be highly unstable and inconsistent with radiation field. To overcome this difficulty, he suggests an alternate use of lambda iteration and linearisation. He also suggests the switching on and off of temperature variation and switching off Newton–Raphson operation before complete convergence. However the points of these switching on and off can only be done from experience. Further, at each depth, he used Ng acceleration with tridiagonal (and pentadiagonal) operators and weights inversely proportional to the variables (Auer (1987)).

Test calculations for WN and WC stars yield favourable results comparable to solutions by methods with diagonal operators.

In the study of line transfer for problems in spherically symmetric moving media, it often happens that spectral lines are formed in sub-sonic photospheres and super-sonic stellar winds. Mihalas *et al.* (1975, 1976) demonstrated that in expanding media, transfer problems are solved effectively in the CMF [cf. chapters 6, 7] particularly in cases of monotonic velocity fields. In problems with super-sonic velocity, the alternative approach of Sobolev approximation [cf. chapter 3] yields good results and saves a fair amount of computer time.

Construction of ALO for Line Transfer Problems in CMF Using Affine Nature of Λ-Operator

Puls (1991) noted that Λ-operator acting on a line source function is of affine type. He utilised this feature to construct a simple, local and optimum ALO for line transfer problem in CMF. This is intimately related to the description of the radiation field in the Sobolev approximation (SA). This ALO is parameter free and purely local. The relation between this local ALO and SA approach allows different possibilities for the formulation of the solution of the rate equations using an arbitrary mixture of Sobolev approximation (SA) and CMF transfer equation. We outline below the basic features of Puls' approach.

In expanding medium, the exact transfer problem is conveniently solved by impact parameter method in CMF as demonstrated in chapters 6 and 7. We recall that the transfer equations in the Feautrier scheme [cf. (6.1.22), (6.1.23)] are written as

$$\frac{\partial u(s,p,\nu)}{\partial \tau(s,p,\nu)} + \gamma(s,p,\nu)\left[\frac{\partial v(s,p,\nu)}{\partial \nu}\right] = v(s,p,\nu), \qquad (8.4.49)$$

and

$$\frac{\partial v(s,p,\nu)}{\partial \tau(s,p,\nu)} + \gamma(s,p,\nu)\left[\frac{\partial u(s,p,\nu)}{\partial \nu}\right] = u(s,p,\nu) - S(s,p,\nu). \qquad (8.4.50)$$

where u and v are mean intensity and flux like variables.

Equations (8.4.49) and (8.4.50) are to be solved under the boundary conditions (6.1.26) and (6.1.27) and the initial condition (6.1.28).

The system of transfer equations, boundary conditions and the initial condition are now discretised. For the solution, Gaussian elimination scheme of the Feautrier or Rybicki type is used at each depth point. We have [cf. (6.1.42)–(6.1.45)]

$$\mathbb{T}_k \mathbf{u}_k = \mathbb{U}_k \mathbf{u}_{k-1} + \mathbb{V}_k \mathbf{v}_{k-1} + \mathbf{S}_k \qquad (8.4.51)$$

$$\mathbf{v}_k = \mathbb{G}_k \mathbf{u}_k + \mathbb{H}_k \mathbf{v}_{k-1}. \qquad (8.4.52)$$

(The above equations have been written in the notation of Puls for easy reference. The term $\mathbb{W}_k \mathbf{J}_k$ is ignored.)

Impact parameter (angle) index 'i' and frequency index "n" are grouped into a single series index k.

We recall that the source function is given by

$$S_k = \frac{\eta_k}{\chi_k} = \frac{\eta_{ck} + \chi_{Lk} S_L}{\chi_{ck} + \chi_{Lk}}. \qquad (8.4.53)$$

At this stage he assumed that the continuum quantities with index ck are constant across the line profile..Integrating over angle and frequency the line transfer mean intensity can be written in terms of an affine operator as

$$\bar{J}_L = \boldsymbol{\Psi}[u_1, v_1, \eta_c, \chi_c, \chi_L] + \Lambda' S_L, \qquad (8.4.54)$$

where $\boldsymbol{\Psi}$ is the displacement vector. In (8.4.53) and (8.4.54) the quantities with subscripts c and L denote those in continuum and line respectively. u_1 and v_1 are blue wing variables refering to the continuum problem This affine equation is linear only when $\eta_c = 0, u_1 = v_1 = 0$. As defined in chapter 6, $\bar{J}_L$ is the frequency integrated mean intensity, which is required for setting up the rate equations.

Construction of Λ-operator

The explicit values of the displacement vector $\boldsymbol{\Psi}$ and the matrix Λ' can be obtained by solving the transfer problem $(D+1)$ times, D being the total number of radial grid points.

We have
$$\boldsymbol{\Psi} = \bar{J}(S_L = 0), \qquad (8.4.55)$$
and
$$\Lambda'_{ij} = \bar{J}_i(S_L = e_j) - \boldsymbol{\Psi}_i, \qquad (8.4.56)$$
where e_j is the unit vector; or
$$\Lambda'_{ij} = \bar{J}_i(S_L = e_j; u_1 = v_1 = \eta_c = 0). \qquad (8.4.57)$$

Accelerated Λ-iteration Cycle

For an affine approximate Λ-operator Λ^A, ALI scheme can be formally written as
$$\bar{J}^n = \Lambda^A [S^n] + \left(\Lambda - \Lambda^A\right) [S^{n-1}]. \qquad (8.4.58)$$

Pauldrach and Herrero (1988) wrote this as
$$\bar{J}^n = \Lambda \left[S^{n-1}\right] + \Lambda^A \left[S^n - S^{n-1}\right]. \qquad (8.4.59)$$

Puls ((1991), p. 583) shows that though the lambda operator is of affine type, only the linear term in it has to be approximated for ALI. For multi-level calculations, he suggested that the lines are iterated with fixed continuum before the continuum is updated again. However, when the population number stabilises and the relative corrections become small, lines and continuum may be iterated in parallel until the desired convergence is achieved.

Approximations of the Diagonal

We have stated earlier that the choice of an ALO of the type [cf. (8.4.5)]

$$\Lambda_\nu^* = diag[\mathbb{T}_\nu^{-1}]$$

is ideal for static problems and ensures a very fast convergence in ALI. $\mathbb{T}_\nu$ is a tridiagonal matrix. However, in CMF the intrinsic coupling of the frequency has to be taken into account when calculating the diagonal elements. Puls ((1991), p. 584) studied the contributions of $u_{k,l}$ (the mean intensity like Feautrier variable) to the exact diagonal of $\Lambda' = \Lambda'_{l,l}$. k is the frequency index for a particular ray. The equation (8.4.57) for $i = j = l$ is to be solved. Equation (8.4.57) now reads

$$\Lambda'_{l,l} = \bar{J}_l(S_L = e_l : u_1 = v_1 = \eta_c = 0).$$

He concluded that for frequencies $k = 1, 2$, $u_{k,l}$ is purely local. For frequencies $k \geq 3$, $u_{k,l}$ are non-local and are coupled with all other $u_{(k-1),j}$ and also with some of the Feautrier's flux like variables. Thus the task of finding the diagonal elements of Λ' is rather involved and time consuming.

Puls instead searched for an appropriate local approximation for $u_{k,l}$. Going back to equation (8.4.51), splitting the operator $\mathbb{V}_k$ into the diagonal $\mathbb{VA}_k$ and $\mathbb{VB}_k$ (with all elements > 0), we get for each k, $k \geq 2$ [cf. Puls (1991), p. 584, equation (9)]

$$\begin{aligned}u_{k,l} &= \sum_i (\mathbb{T}_k^{-1})_{li} \left[\mathbb{U}_{ki} u_{(k-1),i} - \mathbb{VA}_{ki} v_{(k-1),(i-1/2)} \right. \\ &\quad \left. + \mathbb{VB}_{ki} v_{(k-1),(i+1/2)} + \delta_{il}\rho_{ki} \right] \\ &= \sum_i (\mathbb{T}_k^{-1})_{li} b_{ki}, \end{aligned} \qquad (8.4.60)$$

where

$$\rho_k = \frac{\chi_{kL}}{(\chi_{ck} + \chi_{kL})}.$$

b_{ki} is the quantity within the square bracket in (8.4.60). Puls' approximation consists in confining ourselves to local terms only, i.e. in b_{ki} we neglect all the terms except $i = l$.

Let u^*_{kl} be the approximated value of u_{kl} defining the new ALO given by

$$\Lambda'^*_{l,l} = \sum w_{p,l} \sum_{k=1}^{K} w_{kl} u^*_{pk,l} \qquad (8.4.61)$$

where w_{pl} are the quadrature weights for the integration.

To secure convergence of the ALI cycle, we require that $\Lambda'^*_{l,l} < \Lambda'_{l,l}$ where $\Lambda'_{l,l}$ is the exact diagonal operator.

In vector notation, exact $u_{k,l}$ is given by

$$\begin{aligned} \mathbf{u}_k = \mathbb{T}_k^{-1} b_k &= \mathbb{T}_k^{-1} \left(b_{k,1} e_1 + \cdots + b_{kD} e_D \right) \\ &= \sum_i b_{ki} \mathbf{t}_k(e_i) \end{aligned} \qquad (8.4.62)$$

where $\mathbf{t}_k(e_i)$ is the solution of

$$\mathbb{T}_k \mathbf{t}_k = e_i. \qquad (8.4.63)$$

The local approximations is

$$u^*_{k,l} = b_{k,l} t_{k,l}(e_l). \qquad (8.4.64)$$

By definition $t_{k,l}(e_l)$ is positive. The condition $u^*_{k,l} < u_{k,l}$ generally requires that

$$b_{k,l} t_{k,l}(e_l) \leq \sum_i b_{k,i} t_{k,l}(e_i), \qquad (8.4.65)$$

implying that

$$\sum_{i \neq l} b_{k,i} \geq 0. \qquad (8.4.66)$$

Following numerical calculations on a number of models, Puls concludes that the relation (8.4.66) holds good widely (except for a small number of red wing frequencies in the wind around the thermal point).

The local approximation $u^*_{k,l}$ is calculated from

$$\begin{aligned} u^*_{k,0} = diag(\mathbb{T}_k^{-1}) \Big[&\mathbb{U}_k u^*_{(k-1),0} - \mathbb{V}\mathbb{A}_k v^*_{(k-1),(-1/2)} \\ &+ \mathbb{V}\mathbb{B}_k v^*_{(k-1),(1/2)} + \rho_{k,0} \Big]. \end{aligned} \qquad (8.4.67)$$

The Feautrier variables $v_{(k-1),(\pm 1/2)}$ are found from (8.4.52) as

$$v^*_{(k-1),(-1/2)} = \mathbb{G}_{(k-1),(-1/2)}(u^*_{(k-1),0} - u_{(k-1),(-1)})$$
$$+ \mathbb{H}_{(k-1),(-1/2)} v^*_{(k-2),(-1/2)} \qquad (8.4.68)$$

and

$$v^*_{(k-1),(1/2)} = \mathbb{G}_{(k-1),(1/2)}(u_{(k-1),1} - u^*_{(k-1),(0)})$$
$$+ \mathbb{H}_{(k-1),(1/2)} v^*_{(k-2),(1/2)}. \qquad (8.4.69)$$

The boundary conditions at the blue and red wings can be written as

$$u^*_{1,0} = v^*_{1,\pm 1/2} = 0. \qquad (8.4.70)$$

From (8.4.60) we have

$$u_{(k-1),(l\pm 1)} = \sum_i (\mathbb{T}_k^{-1})_{(l\pm 1),i} b_{(k-1),i}. \qquad (8.4.71)$$

From (8.4.67)–(8.4.69), we see that though $u^*_{k,0}$ is written in local approximation, non-local dependence creeps into it through v^* terms defined in (8.4.68) and (8.4.69). This problem could have been dealt with in the same way as above, i.e. by neglecting the contributions of $b_{(k-1),i}, i \neq l$ compared to $b_{(k-1),l}$. Then we have from (8.4.71)

$$u^*_{(k-1),(l\pm 1)} = (\mathbb{T}_{k-1}^{-1})_{(l\pm 1),l} b_{(k-1),l}. \qquad (8.4.72)$$

However, Puls chose a different approach for computational convenience. He neglects the contribution of the off-diagonal elements $u_{(k-1),(l\pm 1)}$ to $v^*_{(k-1),(l\pm 1/2)}$.

Now writing the quantities in the second approximation by "$**$", we have

$$v^{**}_{(k-1),(l-1/2)} = \mathbb{G}_{(k-1),(l-1/2)} u^{**}_{(k-1),l} + \mathbb{H}_{(k-1),(l-1/2)} v^{**}_{(k-2),(l-1/2)} \qquad (8.4.73)$$

$$v^{**}_{(k-1),(l+1/2)} = \mathbb{G}_{(k-1),(l+1/2)} u^{**}_{(k-1),l} + \mathbb{H}_{(k-1),(l+1/2)} v^{**}_{(k-2),(l+1/2)}. \qquad (8.4.74)$$

For the second approximation, in(8.4.69) quantities with superscripts "*" are replaced by "**". With these changes we find that

(a) The approximate diagonal operator $\Lambda_{l,l}^{\prime**}$ depends on purely local quantities $u_{k,l}^{**}$, $v_{k,(l\pm 1/2)}^{**}$, for all frequencies.

(b) $\quad \Lambda_{l,l}^{\prime**} \leq \Lambda_{l,l}^{\prime*} \leq \Lambda_{l,l}^{\prime}$ \hfill (8.4.75)

(c) in the static limit $\Lambda_{l,l}^{\prime} = \Lambda_{l,l}^{\prime**}$ \hfill (8.4.76)

Puls (1991) suggests the following routine for constructing his approximate Λ^{**}.

For each ray

(a) one starts with $u_{1,0}^{**} = v_{1,\pm 1/2}^{**} = 0$,

(b) bracketed term on the right hand side of equation (8.4.69) is calculated for each frequency k ($k = 2, 3, \ldots, K$) from prior solution of $u_{(k-1),0}^{**}$ and $v_{(k-1),\pm 1/2}^{**}$

(c) $\mathbb{T}_k^{-1}$ is calculated by Rybicki and Hummer (1990) algorithm and $u_{k,0}^{**}$ obtained from modified (8.4.69) and $v_{k,\pm 1/2}^{**}$ from (8.4.73) and (8.4.74),

(d) updated values of vectors $u_{(k-1),0}^{**}$ and $v_{(k-1),0}^{**}$ are added to $\Lambda^{\prime**}$ with appropriate spatial and frequency weights,

(e) loop over all frequencies for all rays are calculated to obtain $\Lambda^{\prime**}$ and $\bar{J}$.

Puls concludes that the constructed $\Lambda^{\prime**}$ is a purely a local operator. It ensures absence of any spatial coupling and corresponds fully to the Sobolev approximation. Secondly the local CMF operator and ALO's by construction reach the static limit for small velocities. Puls ((1991), p. 586–587] examined the use of ALI type [cf. (8.4.58), (8,4,59)] solutions for rate equations as well. He found that the formal correspondence between the local CMF and Sobolev approximation (SA) approaches suggests a number of computationally economical methods of solving the rate equations. The correspondence between CMF and SA approaches opens the way of mixing them judiciously to secure the fast solutions of the coupled equations of transfer and statistical equilibrium. He recommends

(a) direct solution of rate equations by Sobolev approximation till convergence,

(b) to decide from it lines to be treated in CMF,

(c) use of SA or local CMF operators for direct solution with a view to obtaining a good approximation to CMF lines,

(d) finally taking recourse of ALI scheme till convergence (using mixed operators).

The method proposed by Puls (1991) for solving line transfer problems in expanding atmospheres is parameter free and works at optimum convergence rate. This method of local approximate lambda operator bears ample resemblance to Sobolev's approximation in terms of basic principles.

Sellmaier, et al. (1993) treated the non-LTE multi-level radiative transfer in unified model (photosphere plus wind) atmosphere in the presence of both sub- and supersonic velocity fields. They used the local approximate lambda operator ALO developed by Puls (1991) and an ALI scheme for solving line transfer problem in CMF. For complex atomic models, the scheme gives excellent convergence. Many strong, weak and intermediate lines can be studied simultaneously. They observed that the method based on core saturation factor [cf (8.4.16)] is inadequate in predicting the line formation in all ranges-strong,weak and intermediate lines.They also noted that CMF-diagonal operator of Puls (1991) is the best local ALO for CMF transport in unified models. This ALO leads to very fast convergence of the line iteration problem even when as many as 600 line transitions are included.

Hamann, Koesterke and Wesselowski (1991) proposed a quasi-Newtonian iterative scheme together with the use of Broyden update formula. It is found to give considerable computational advantage. The repeated calculations of derivatives and inversion of matrices can be avoided here.

Koter, Schmutz and Lamers (1993) suggested a fast non-LTE code to calculate the continuum energy distribution and line profiles in an expanding atmosphere. They used a semi-empirical

8.4 Operator Perturbation Method

model of expanding atmosphere with chosen density, velocity and temperature structures. The statistical equilibrium and radiative transfer equations in the continuum are solved with ALI. The line transfer is treated with Sobolev approximation.

It has already been mentioned that the statistical equilibrium equations are implicitly non-linear in the population numbers in levels through the influence of the radiation field. In ALI [cf. (8.4.4)] the rate equations become explicitly non-linear in level population numbers through the introduction of S^{new}. Koter et al. preconditioned the photo-ionisation and stimulated recombination terms in the statistical equilibrium equations in such a way that the rate equations remain linear. The need for this preconditioning process was first felt by Rybicki (1972). Pauldrach and Herrero (1988) and Rybicki and Hummer (1991) examined the preconditioning of line transfer in terms of ALI scheme to linearise the continuum problem. Koter et al. [(1993), p. 564] used it for their problem. The Sobolev approximation used by them includes absorption and emission processes in the extensive atmosphere and continuous absorption within the resonance zone. They termed this method ISA standing for Improved Sobolev Approximation. For solving non-LTE radiative transfer problems in spherically symmetric expanding atmospheres under constraints of statistical equilibrium, the most contemporary and effective method used is the ALI approach. A comprehensive review of the general scheme of these methods have been given by Hubeny (1992) and Rybicki (1991) We have outlined above some examples of the use of the ALI approach for solving coupled equations of transfer and rate equations relevant to the line transfer problems in spherically symmetric expanding media. Efforts are continuing in this direction for developing effective schemes for explaining the spectral structure of special lines in special class of stars. Recently Hillier (1996) has used the ALI approach to study the line polarisation in axi-symmetric envelopes.

The basic mathematical tools used in the ALI method may be summarised as follows:

(a) Auer and Mihalas's (1969) complete linearisation scheme for radiative transfer and constraint equations.

(b) Extension of Cannon's (1973a,1973b) and Scharmer's (1981, 1984) operator perturbation scheme to spherically symmetric expanding atmospheres. Their scheme consists in simplified description of radiative transfer through a simple (mostly local) approximate integral operator (ALO), linearising the source function about an initial value and developing an iteration scheme to correct the error incurred due to the use of ALO instead of the exact integral operator. The elements of core saturation method of Rybicki (1972,1984) and Newton-Raphson linearisation scheme is widely used to solve the non-linear problems.

(c) Wide use is made of Feautrier's impact parameter approach for solving transfer problems in spherically symmetric moving media using CMF and Sobolev's approximation (SA) or a mixture of both.

(d) ALI process is developed for rapid convergence. Ng's (1974) and Auer's (1987) acceleration schemes are sometimes used for faster convergence.

In all the ongoing works of the subject, some of the above mentioned methodologies have been mixed judiciously and often used alternatively to solve these problems. Their main goal is to secure fast convergence of computation without adversely sacrificing the accuracy of the solutions.

ALO-ALI Technique for Computing $\triangle I_i^{\pm}$ in the Short Characteristic Rays Method

The basic structure of ALO–ALI technique and their uses have been detailed in section (8.4). We briefly outline below the process of evaluating $\triangle I_i^{\pm}$ by ALO–ALI scheme. Before we do this let us first view the integral operator on $\hat{S}(t)$ occurring in equation (8.3.10) as a matrix operator on the vector $(\hat{S}_1, \ldots, \hat{S}_N)$. Here $\hat{S}_i$ are the discretized values of $\hat{S}(t)$ at the spatial grid points r_i. With this view point, we are in a position to define the approximate operator Λ^*. We choose the matrix representing Λ^* to be close to the matrix Λ but retain only the diagonal terms or at most a few off diagonal items. Hauschildt has suggested using the tridiagonal

8.4 Operator Perturbation Method

matrix for better convergence but in the present discussion we will restrict to obtaining only the relevant diagonal elements of Λ^*. As mentioned earlier we have to distinguish between disc and shell rays in spherical atmospheres. Let us write

$$\Lambda^* = \Lambda^c + \Lambda^t$$

where Λ^c represents the core intersecting disc ray and Λ^t represents the shell ray that is tangent to the radial grid.

In calculating the ith column of the approximate lambda matrix Λ^*, we set $\hat{S}_i = 1$, $\hat{S}_{j \neq i} = 0$; take the incident intensities at the boundaries to be zero and obtain a formal solution. The result will be a normalized intensity which when integrated over frequencies and characteristic rays will give the desired columns of Λ^*. This is strictly not correct as $\hat{S}$ has contributions not only from S but also from the $\partial/\partial \lambda$ term. None the less this approximation is used to save the computational time. Hauschildt (1992) has also discussed ways of improving this approximation by incorporating the effects of the $\partial/\partial \lambda$ term. We will not pursue this but take the original approximation and illustrate how to compute the diagonal elements of Λ^*.

Let us first consider the core intersecting rays and compute Λ^c. From our above assumptions, the inward directed normalized intensities will remain zero until grid point $i - 1$. Hence

$$\hat{I}^-_{i,j}(\nu) = -\beta^-_{i,j}$$
$$\hat{I}^-_{i+1,j}(\nu) = \beta^-_{i,j} \exp(\triangle \tau_{i,j}) + \alpha^-_{i+1,j}$$
$$\hat{I}^-_{k,j}(\nu) = \hat{I}^-_{k-1,j}(\nu) \exp(-\triangle \tau_{k-1,j}), \quad k = i+2, \ldots, N.$$

Also the outward directed normalized intensity will remain zero until the grid $(i + 1)$. Hence

$$\hat{I}^+_{i,j}(\nu) = \beta^+_{i,j}$$
$$\hat{I}^+_{i-1,j}(\nu) = \beta^+_{i,j} \exp(\triangle \tau_{i-1}) + \gamma^+_{i-1,j}$$
$$\hat{I}^+_{k,j}(\nu) = \hat{I}^+_{k+1,j} \exp(-\triangle \tau_{k,j}), \quad k = i-2, i-3, \ldots, 1.$$

The diagonal elements Λ^c_{ii} are given by

$$\Lambda^c_{ii} = \int_{-\infty}^{\infty} \phi d\nu \left[\sum_j \hat{I}^-_{i,j} + \hat{I}^+_{i,j} \right] = \sum_k \sum_j w_{k,j} \left(\hat{I}^-_{i,j} + \hat{I}^+_{i,j} \right),$$

where $w_{k,j}$ are the weights for the wavelength-angle quadrature.

To obtain the diagonal elements Λ^t we have to consider multiple intersections of tangent rays with some shells. [cf. Hauschildt (1992) equations (22)–(32)]. For instance the outward directed intensity ($\mu > 0$) can be computed as follows. Since there is a source $S_i = 1$ at r_i, there will be contributions to the normalized outward intensity from the source at r_i from both the points r_i with $s > 0$ and $s < 0$. For $i > j$, let

$$\hat{I}_{i,j}^{+(1)}(\nu) \mathrel{+}= \hat{I}_{\ell,j}^{-}(\nu), \quad \ell = i, i+1, \ldots, j+1;$$

$$\hat{I}_{i,j}^{+(2)}(\nu) \mathrel{+}= \hat{I}_{\ell,j}^{+}(\nu), \quad \ell = j, \ldots, i;$$

with

$$\hat{I}_{j,j}^{+}(\nu) = \hat{I}_{i,j}^{+(1)}(\nu).$$

Here the symbol "+=" denotes the do-loop summation in standard programming language.

The above normalized intensity is the contribution from the source at r_i with $s < 0$. The total normalized outward directed intensity is given by

$$\hat{I}_{i,j}^{+(n)}(\nu) = \hat{I}_{i,j}^{+}(\nu) + \hat{I}_{i,j}^{+(2)}(\nu).$$

Similarly the inward directed intensity ($\mu < 0$) can be computed and hence the element Λ_{ii}^t can be obtained. It must be recalled that the optical depths along the rays between the spatial grid points are calculated by assuming piecewise linear interpolation of the extinction coefficients. To ensure good accurary in the μ discretzation, Hauschildt (1992) has suggested taking a few extra core intersecting characteristic rays in addition to the rays tangent to the spatial grids. The scheme, in addition to being fast, gives accurate results.

References

Auer, L.H. (1987): Numerical Radiative Transfer, Cambridge University Press, Cambridge. p. 101.
Auer, L.H., Mihalas, D. (1978): Ap. J., **158**, 641.
Cannon, C.J. (1973a) : JQSRT, **13**, 627.
Cannon, C.J. (1973b) : Ap. J., **185**, 621.
Chandrasekhar, S. (1960): Radiative Transfer, Clarendon Press, Oxford.

References

Grant, I.P., Hunt, G.E. (1968): Mon. Not. R. Ast. Soc., **141**, 27.
Grant, I.P., Hunt, G.E. (1969a): Proc. R. Soc. Lond. A, **313**, 183.
Grant, I.P., Hunt, G.E. (1969b): Proc. R. Soc. Lond. A, **313**, 199.
Hamann, W.R. (1987): Numerical Radiative Transfer, ed. Kalkofen. W. Cambridge University Press, Cambridge. p. 35.
Hamann, W.R. (1985): Astron. Astrophys., **148**, 364.
Hamann, W.R. (1986): Astron. Astrophys., **160**, 347.
Hamann, W.R., Koesterke, L., Wessolowski, U. (1991): Stellar Atmosphere; Beyond Classical Models, NATO ASI Ser. C 341, ed. Crivellari, L., Hubeny, I., Hummer D.G. Kluwer Publications.
Hamann, W.R., Wessolowski, U. (1990): Astron. Astrophys., **227**, 17.
Hauschildt, P.H. (1992): JQSRT, **47**, 433.
Hauschildt, P.H., Wehrse, R. (1991): JQSRT, **41**, 81.
Hempe, K., Wessolowski, U. (1986): Astron. Astrophys., **160**, 141.
Hillier, D.J., (1990): Astron. Astrophys., **231**, 116.
Hillier, D.J. (1996): Astron. Astrophys., **308**, 521.
Hubeny ,I. (1992): The Atmospheres of early type of stars.ed. Herber, V., Jeffrey, C. Lecture Notes in Physics, 401, Springer-Verlag, Berlin.
Kalkofen, W. (1984): Methods in Radiative Transfer, Cambridge University Press, Cambridge.
Kalkofen, W. (1974): Ap. J., **188**, 105.
Kalkofen, W. (1987): Numerical Radiative Transfer, Cambridge University Press, Cambridge.
Koter, de., Schmutz, W., Lamers, H.J.G.L.M. (1993): Astron. Astrophys., **277**, 561.
Kunasz, P.B., Auer, L.H. (1987): JQRST, **39**, 67.
Kourganoff, V. (1963): Basic Methods in Transfer Problems, Dover Publications, Inc., New York.
Mihalas, D. (1978): Stellar Atmospheres, W.H. Freema and Co., New York.
Mihalas, D. (1980): Ap. J., **237**, 574.
Mihalas, D., Mihalas, B. W. (1984a): Foundations of Radiation Hydrodynamics, Oxford University Press, New York.
Mihalas, D., Winkler, K.H.A., Norman, M.L. (1984b): JQRST, **31**, 473.
Mihalas, D., Winkler, K.H.A., Norman, M.L. (1984c): JQRST, **31**, 479.
Ng, K.C. (1974): J. Chem. Phys. **61**, 2680.
Olson, G.L., Auer, L.H., Buchler, J.R. (1986): JQSRT, **35**, 43.
Olson, G.L., Kunasz, P.B. (1987): JQSRT, **38**, 325.
Pauldrach, A., Herrero, A. (1988) : Astron. Astrophys. **199**, 262.
Puls, J. (1991) : Astron. Astrophys., **248**, 581.
Puls, J., Herrero, A. (1988) : Astron. Astrophys., **204**, 219.
Peraiah, A. (1991a): Ap. J., **371**, 673.
Peraiah, A.(1991b): Ap. J., **380**, 212.
Peraiah, A., Grant, I.P. (1973): J. Inst. Math. Appl. **12**, 75.
Peraiah, A., Varghese, B.A. (1985): Ap. J. **290**, 411.
Rybicki, G.B. (1971) : JQSRT, **11**, 589.
Rybicki, G.B. (1972) : Line formation in the presence of magnetic fields, eds.Athay,R.G.,House,F.B., Newkirk,G.Jr. High Altitude Observatory Boulder, p. 145.
Rybicki, G.B. (1984): Methods in Radiative Transfer, ed. Kalkofen, W. Cambridge University Press, Cambridge. p. 21.
Rybicki, G.B., Hummer D.G. (1991): Astron. Astrophys., **245**, 171.
Rybicki, G.B. (1991): Stellar Atmosphere; Beyond Classical Models, NATO ASI Ser. C 341, eds. Crivellari, L., Hubeny, I., Hummer, D.G. Kluwer Publications, Dordrechet.

Scharmer, G.B. (1981): Ap. J. **249**, 720.
Scharmer, G.B. (1984): Methods in Radiative Transfer, ed. Kalkofen, W. CambridgeUniversity Press, Cambridge, p. 173.
Scharmer, G.B., Carlsson, M. (1985): J. Comp. Phys., **59**, 56.
Scharmer, G.B., Nordlund, A. (1982): Stockholm Obs. Rep., **19**.
Schmutz, W., Hamann, W. R., Wessolowski, U. (1989) : Astron. Astrophys., **210**, 236.
Sellmaier, F., Puls, J., Kudritzki, R. P., Gabler, A., Gabler, R., Vouls, S. A., (1993): Astron. Astrophys., **273**, 533.
Schönberg, K., Hempe, K. (1986): Astron. Astrophys., **163**, 151.
Sobolev, V.V. (1960): Moving Envelopes of stars, Harvard University Press.
Werner, K. (1986): Astron. Astrophys., **161**, 177.
Werner, K. (1987): Numerical Radiative Transfer, ed. Kalkofen, W. Cambridge University Press, Cambridge, p. 67.
Werner, K., Husfield, D. (1985): Astron. Astrophys., **148**, 417.
Winkler, K.H.A., Norman, M.L., Mihalas, D. (1984): JQSRT, **31**, 473.
Winkler, K.H.A., Mihalas, D., Norman, M.L. (1985): Computer Physics Communications, **36**, 121.
Winkler, K.H.A., Norman, M.L. (1983): Astrophysical Radiation Hydrodynamics, D. Reidel Publishing Co., Dordrechet.

Glossary of Physical Symbols

a	:	acceleration
$\tilde{a}$	:	$\frac{\nu V(r)}{cr}$
A_{ki}	:	Einstein coefficient for spontaneous emission
A_{ci}	:	number of recombinations onto level i
$A_{\ell u}$	:	Einstein coefficient for $l \to u$ transfer for two-level atoms
$\mathbb{A}_{(d+1/2)}$	:	matrix associated with impact parameter method
$B(\nu, T) = B_\nu(T)$	:	Planck function
$B(T)$	:	frequency integrated Planck function
B_{ki}	:	Einstein coefficient for stimulated emission
$B_{\ell u}, B_{u\ell}$	:	Einstein coefficients for transfer between lower and upper states in two level atoms
$\mathbb{B}_{d+1/2}$	:	matrix associated with impact parameter method
c	:	velocity of light
C_{ki}	:	the number of collision induced transition $k \to i$
C_{ci}	:	number of three body recombinations into level i
C	:	continuum state
$C_{u\ell}, C_{\ell u}$	:	collision rates between upper and lower levels in two-level atom
C_{ij}	:	scattering cross-section
$\mathbb{C}_{d+1/2}$	:	matrix associated with impact parameter method
$d\Sigma$	:	infinitesional area
$d\Omega$	:	element of solid angle

$d\nu$	:	frequency interval
ds	:	elementary length along the radiation pencil
$\mathbb{D}_{d+1/2}$	:	matrix associated with impact parameter method
E	:	radiant energy
E	:	photon energy $= h\nu$
E_R	:	energy in the radiation field
e	:	electronic charge
$f_R(\mathbf{r}, t; \hat{\mathbf{s}}, \nu)$	:	photon distribution function
$\bar{F}_\nu$	:	monochromatic flux
F_ν	:	$\bar{F}_\nu = \pi F_\nu$
F	:	frequency integrated F_ν
f_{ij}	:	oscillation strength for $i \to j$ transition
f_ν	:	variable Eddington factor, $f_\nu = K_\nu / J_\nu$
g_i, g_j	:	statistical weights of atomic states i and j
h	:	Planck's constant
H_ν	:	$H_\nu = \frac{1}{4}(F_\nu)$, first angular moment
H	:	frequency integrated H_ν
h_ν	:	$H_\nu(R)/J_\nu(R)$ defined in (4.2.13)
$I(\mathbf{r}, t; \hat{\mathbf{s}}, \nu)$	:	specific intensity
$I(\mathbf{r}, \mu, \nu)$	:	specific intensity
$I(\mathbf{r}, \hat{s})$	:	specific intensity
$I_0(s, p, \nu)$	:	specific intensity
$I(r_0, \mu_0, \nu_0)$	:	specific intensity in comoving frame
i	:	imaginary sign $i^2 = -1$
I_c	:	intensity from the core source
$I^+(r, \mu, \nu)$	:	$I^+(r, \mu, \nu) = I(r, +\mu, \nu), 0 \leq \mu \leq 1$
$I^-(r, \mu, \nu)$	:	$I^-(r, \mu, \nu) = I^-(r, -\mu, \nu), 0 \leq \mu \leq 1$
$\mathbb{I}$	:	identity matrix
$J(\mathbf{r}, t, \nu)$	:	mean intensity; zeroth angular moment

$J(\mathbf{r},\nu)$	:	mean intensity; zeroth angular moment of specific intensity
$J_\nu(\mathbf{r})$	:	mean intensity; zeroth angular moment of specific intensity
$\bar{J}$	:	frequency integrated mean intensity
$\bar{J}_{ik}$	:	mean intensity in the $i \to k$ line
K_ν	:	second angular moment of specific intensity
K	:	frequency-integrated K_ν
$k(\mathbf{r},t;\hat{s},\nu)$	:	absorption coefficient
$k_\nu(r,t)$	:	absorption coefficient
$k(r)$	:	absorption coefficient
$K(r,r')$	:	kernel function
l	:	characteristic length of material
L_ν	:	third angular moment of specific intensity in lab frame
L	:	Sobolev length
L	:	Lorentz transformation
L_{lm}	:	luminosity
m_0	:	rest mass
M	:	fourth angular moment of specific intensity in lab frame
$\mathbf{M}_k$	:	a vector containing depth variation of thermal term
n_k	:	population at level k
n_e	:	electron density
N	:	fifth angular moment of specific intensity
$P^{ij}(\mathbf{r},t;\nu)$	:	component of radiation pressure tensor
$\mathcal{P}_\nu = \mathcal{P}(\mathbf{r},t,\nu)$	:	radiation pressure tensor
P	:	radiation pressure
$p(\hat{s},\hat{s}')$	:	phase function
$p^0(\mu,\mu')$	:	phase function (isotropic scattering case)
p	:	photon momentum $= h\nu/c$
p_ν	:	monochromatic photon escape probability
p_e	:	photon escape probability parameter
p_c	:	photon escape probability for photon emitted from source to the test point
$Q(r,\mu)$	:	$\mu^2 \frac{\partial v}{\partial r} + (1-\mu^2)\frac{V}{r}$

q_ν	:	sphericity factor defined by (4.2.7)
(r, θ, φ)	:	polar coordinates
$R(s', \nu'; \hat{s}, \nu)$	:	redistribution function
$R(\nu', \nu)$	:	redistribution function
R	:	outer radius of the spherical atmosphere
$\hat{s}$	:	unit vector in the direction of ray
$S(\mathbf{r}, \nu)$	:	source function
t_f	:	fluid flow time
t_R	:	radiation flow time
t_d	:	radiation diffusion time
$\mathbb{T}_k$	:	matrix associated with impact parameter method
$u_{\nu_{jc}}$	:	Planckian radiation density of frequency
u	:	$u = \frac{1}{2}[I^+(r,\mu,\nu) + I^-(r,\mu,\nu)]$
$\mathbb{U}_k$	:	diagonal matrix containing depth variation
$V(r)$	:	velocity of the fluid frame
v_{th}	:	thermal velocity
v	:	$v = \frac{1}{2}[I^+(r,\mu,\nu) - I^-(r,\mu,\nu)]$, also velocity
v_b	:	line broadering velocity
$\mathbb{V}_k$	:	matrix associated with impact parameter method
w	:	dilution factor
w_k	:	quadrature weight for combined angle-frequency quadrature
x^1, x^2, x^3, x^4	:	four coordinates in lab frame
$x_0^1, x_0^2, x_0^3, x_0^4$	:	four coordinates in comoving frame
$X_{d,i,n}$	:	a variable $X_{din} = X(\tau_d, \mu_i, \nu_n)$
x	:	$x = (\nu - \nu_0)/\triangle\nu_D$
$\alpha_{(d+1/2),n}$	:	discrete quantity connected with scattering coefficient at depth points $(d+1/2)$ and frequency point n
$\wedge$	:	lambda operator
$\wedge^*$	:	approximate lambda operator
β	:	$\beta = v/c$
$\beta_{(d+1/2),k}$	:	source function at discrete depth-angle-frequency point

$\triangle \nu_D$	:	Doppler width
γ	:	$\gamma = 1/(1-\beta^2)^{1/2}$
ε	:	fraction of line emission of thermal origin
ε'	:	$\varepsilon = \varepsilon'/(1+\varepsilon')$
η	:	total emissivity
η^T	:	true or thermal emission coefficient
η^S	:	emission coefficient due to scattering
η^0	:	emissivity in comoving frame
μ, μ'	:	$\cos\theta$ (cosine of polar angle of raidation pencil)
μ^0, μ_0'	:	$\cos\theta^0, \cos\theta'$ measured in comoving frame
μ_i	:	discrete angle-points, in depth-angle-frequency mesh, $(i=1,2,\ldots,I), 0 \leq \mu \leq 1$
λ_p	:	photon mean free path
ν_0	:	line centre frequency
ν_0	:	frequency in comoving frame
ν	:	frequency in lab frame
ν'	:	frequency in the moving frame
ν_n	:	discrete frequency-points in depth-angle-frequency mesh $0 < \nu < \infty$
τ	:	optical depth, optical thickness
τ_d	:	discrete optical depth points in depth-angle-frequency mesh $(d=1,2,\ldots,D)$
$\chi(\mathbf{r},t;\hat{\mathbf{s}},\nu)$	:	attenuation or extinction
$\chi(\mathbf{r},\nu)$	:	attenuation coefficient
χ^0	:	attenuation coefficient measured in comoving frame
φ	:	azimuthal angle
$\phi(\nu)$	:	line absorption profile
$\Phi(\nu)$	:	line profile in moving media
ψ	:	line emission profile
$\omega(r,\nu)$	:	albedo for single scattering
$\omega(t)$	:	$\omega(t) = \int_{-\infty}^{t} \phi(t')dt'$.

Index

Aberration 8, 41
  effect 51, 158
Absorption
  profile 9
Absorption
  coefficient 7
Advection 8, 41
  effect 158
Albedo 14
Attenuation coefficient 12

Boltzmann
  equation 2
  constant 14
Broyden formula 234

Characteristics 159, 177
  equations 207
  short 206
Core fraction 217
Core saturation method 140

Derivative
  adaptive-mesh 200
  Eulerian 200
  Lagrangian 70, 200
Detailed balancing 37
Diffusion approximation 117, 171
Dilution factor 88
Direct quadrature method 129
Discrete space scheme 121
Doppler
  effect 42, 124
  profile 132
  shift 51, 148
  width 124, 132

Eddington
  factor 111, 117
  generalised relations 184
  approximation 6
  factor 6, 167, 171
Eddington
  factor 115
  generalised approximation 183
Einstein coefficient 32, 91
Elimination scheme 156, 172
Emission
  coefficient 16
  profile 10
Emissivity
  continuum 150
  line 150
Escape probability 91
  parameter 80
Escape probability
  function 138
Extinction coefficient 12

Feautrier elimination scheme 109
Feautrier scheme 108, 151
Frame
  comoving 42
  lab 42
  mixed 42

Impact parameter 104, 118, 147
  Rybicki's modification 112
  spherical moving media 122
Intensity
  angular moments 166, 175
Interaction principle 190
Interpolating multinomials 194

Kirchoff's law 17

Lambda iteration
  accelerated 213, 229
  approximate 214
  classical 213
  improved 235
Lambda operator

approximate 212, 214
Line
  profile 12, 13
  profile 89
Line
  absorption coefficient 35
Linearisation scheme 212
Lorentz transformation 48
  mixed frame 58
  radiation related variables 49
  transfer equation 54
LTE 17, 22

Mean
  absorption 205
  flux 205
  Planck 205
Mean intensity 10, 23
Milne problem 22
Moment equations
  comoving frame 67
  lab frame 43
  mixed frame 57
  mixedframe 58
Monochromatic
  flux 5
  optical depth 138
  radiant energy 4
Monochromatic radiation
  energy equations 67
  momentum equations 68
  energy equation 44
  energy equations 58
  momentum equations 44, 59

Newton-Raphson iteration 221
Ng's acceleration 227
Ng's acceleration 222
Non-LTE 30

Opacity 12
  continuum 150
  line 92, 150
Operator perturbation 211
Optical depth 21, 130
Optical thickness 84
Oscillation strength 86

Phase function 11
Photon distribution function 2
Planck
  constant 14
  function 14
Planckian radiation density 37

Pressure tensor 6, 56
Probability methods
  first order 98, 139
  second order 98
Probability of penetration 87

Radial function 198
Radiative and collisional transitions 30
Rate equations 31
Redistribution function 8
  complete 12
Reflection operators 190
Resonant condition 94
Reynolds theorem 200
Rosseland cycle 100

Saturated core region 141
Scattering cross section 86
Sobolev
  approximation 227
  escape probability method 77
  generalised method 92
  length 81
Source function
  comoving 228
  comoving frame 150, 182
  lab frame 14, 18, 126
Spectral lines 75
Sphericity factor 116
Star products 189, 191
Statistical equations 30, 31, 35
Structure function 198

Transfer equation
  conservative form 202
  adaptive mesh 202
  comoving frame 65
  spherical comoving frame 47
  spherical geometry 26
  stationary medium 16
Transmission operators 190
Two-level model 88, 139